CONTENTS

DECEMBER 4, 2001

OPENING STATEMENTS

	Page
Carper, Hon. Thomas R., U.S. Senator from the State of Delaware	4
Clinton, Hon. Hillary Rodham, U.S. Senator from the State of New York	15
Corzine, Hon. Jon S., U.S. Senator from the State of New Jersey	11
Inhofe, Hon. James M., U.S. Senator from the State of Oklahoma	4
Jeffords, Hon. James M., U.S. Senator from the State of Vermont	1
Smith, Hon. Bob, U.S. Senator from the State of New Hampshire	2
Voinovich, Hon. George V., U.S. Senator from the State of Ohio	13, 18

WITNESSES

Grosser, Mike, technical director, Nuclear Biologic and Chemical Defense Systems, Marine Corps Systems Command	26
Prepared statement	52
Meehan, Patrick, M.D., Director, Division of Emergency and Environmental Health Services, National Center for Environmental Health, Centers for Disease Control and Prevention, Department of Health and Human Services	21
Prepared statement	46
Responses to additional questions from:	
Senator Corzine	49
Senator Jeffords	48
Senator Smith	49
Vinney, Les, president and CEO, STERIS Corporation accompanied by: Peter Burke, vice president and chief technology officer; Gerry Reis, vice president for Corporate Administration; Karla Perri, senior environmental consultant, Versar, Inc.	28
Comments on the Proposed Remediation Plan for the Hart Senate Building (HSOB)	64
Comparison of Antimicrobial Foam, Liquid Chlorine Dioxide and Registered Sporicidal Products	67
Comparison of Antimicrobial Fogging, Chlorine Dioxide and VHP for Room Decontamination	66
Comparison of Bleach and Registered Sporicidal Products for Liquid Surface Disinfection	67
Comparison of Room Decontamination Methods	68
Detailed Biological Remediation Plan	58
Features of Current Proposal and Proposed Alternatives	67
Overview	57
Prepared statement	55
Responses to additional questions from:	
Senator Jeffords	62
Senator Smith	63
Walks, Ivan, M.D., director, District of Columbia Department of Health	23
Prepared statement	51
Whitman, Hon. Christine Todd, Administrator, Environmental Protection Agency	5
Prepared statement	34
Responses to additional questions from:	
Senator Corzine	39
Senator Jeffords	37

Page

Whitman, Hon. Christine Todd, Administrator, Environmental Protection Agency—Continued
—Continued
Senator Smith 40

ADDITIONAL MATERIAL

Comments on the Proposed Remediation Plan for the Hart Senate Building (HSOB) 64
Comparison of Antimicrobial Foam, Liquid Chlorine Dioxide and Registered Sporicidal Products 67
Comparison of Antimicrobial Fogging, Chlorine Dioxide and VHP for Room Decontamination 66
Comparison of Bleach and Registered Sporicidal Products for Liquid Surface Disinfection 67
Comparison of Room Decontamination Methods 68
Detailed Biological Remediation Plan 58
Features of Current Proposal and Proposed Alternatives 67
Letter, Battelle Memorial Institute 69
Summary, Decontamination Methods—Bacterial Spores, Battelle Memorial Institute 70
Tables, Remediation plans for Federal Buildings 66–69

S. Hrg. 107–665

REMEDIATION OF BIOLOGICALLY AND CHEMICALLY CONTAMINATED BUILDINGS

HEARING

BEFORE THE

COMMITTEE ON ENVIRONMENT AND PUBLIC WORKS UNITED STATES SENATE

ONE HUNDRED SEVENTH CONGRESS

FIRST SESSION

ON

REVIEW OF CLEANUP ACTIVITIES IN FEDERAL BUILDINGS AFFECTED BY ANTHRAX CONTAMINATION

DECEMBER 4, 2001

Printed for the use of the Committee on Environment and Public Works

U.S. GOVERNMENT PRINTING OFFICE

81–720 PDF WASHINGTON : 2003

For sale by the Superintendent of Documents, U.S. Government Printing Office
Internet: bookstore.gpo.gov Phone: toll free (866) 512–1800; DC area (202) 512–1800
Fax: (202) 512–2250 Mail: Stop SSOP, Washington, DC 20402–0001

COMMITTEE ON ENVIRONMENT AND PUBLIC WORKS

ONE HUNDRED SEVENTH CONGRESS
FIRST SESSION

JAMES M. JEFFORDS, Vermont, *Chairman*

MAX BAUCUS, Montana
HARRY REID, Nevada
BOB GRAHAM, Florida
JOSEPH I. LIEBERMAN, Connecticut
BARBARA BOXER, California
RON WYDEN, Oregon
THOMAS R. CARPER, Delaware
HILLARY RODHAM CLINTON, New York
JON S. CORZINE, New Jersey
BOB SMITH, New Hampshire
JOHN W. WARNER, Virginia
JAMES M. INHOFE, Oklahoma
CHRISTOPHER S. BOND, Missouri
GEORGE V. VOINOVICH, Ohio
MICHAEL D. CRAPO, Idaho
LINCOLN CHAFEE, Rhode Island
ARLEN SPECTER, Pennsylvania
BEN NIGHTHORSE CAMPBELL, Colorado

KEN CONNOLLY, *Majority Staff Director*
DAVE CONOVER, *Minority Staff Director*

REMEDIATION OF BIOLOGICALLY AND CHEMICALLY CONTAMINATED BUILDINGS

TUESDAY, DECEMBER 4, 2001

U.S. SENATE,
COMMITTEE ON ENVIRONMENT AND PUBLIC WORKS,
Washington, DC.

The committee met, pursuant to notice, at 9:30 a.m. in room 406, Senate Dirksen Building, Hon. James M. Jeffords (chairman of the committee) presiding.

Present: Senators Jeffords, Smith, Carper, Voinovich, Corzine, and Clinton.

OPENING STATEMENT OF HON. JAMES M. JEFFORDS, U.S. SENATOR FROM THE STATE OF VERMONT

Senator JEFFORDS. The hearing will come to order.

I'd like to begin by thanking all of our witnesses for participating in today's hearing on building decontamination.

Last week, several congressional committees held hearings on various aspects of our experience with bioterrorism. Concerns were raised about the ability to quickly and efficiently respond with appropriate assessment and decontamination protocols. A resonating conclusion was a need for more consistent information and more centralized leadership. One position stated that we have been operating in an informational vacuum. I agreed and that is why we are here today. Sadly, terrorism has become a fact of life and although our law enforcement officials are working diligently to ensure our safety, we must take every measure to guarantee our preparedness. That means that we need to learn quickly from our current difficulties.

We are here today seeking knowledge in three areas: first, the coordination that goes into decontaminating a building; second, the health aspects of both cleanup technologies and residual contaminants; and finally, the various technologies available for remediating a building.

I believe that this hearing is critical as a forum in which we can all learn. After all, we are the test case and photos you will see here today document a historical event. Affected parties such as the U.S. Postal Service are awaiting a decontamination model to emerge out of EPA's current efforts to remediate the Hart Senate Office Building. No prior attempt has ever been made to remediate a biologically contaminated building. In fact, 2½ years ago, the Working Group on Civilian Biodefense published a report which stated that such a decontamination effort would be extremely difficult. Well, leave it to Congress to expect to rewrite the science. Here we are.

EPA has been given a tremendous responsibility despite the lack of prior experience and systematic protocol. Therefore, I am pleased that we have Governor Christine Todd Whitman with us, the EPA Administrator, and I would like to thank her for her current efforts and offer my assistance as we both learn about the response protocol necessary to effectively address acts of bioterrorism.

I turn to my compatriot here.

The prepared statement of Senator Jefford follows:]

STATEMENT OF HON. JAMES M. JEFFORDS, U.S. SENATOR FROM THE STATE OF VERMONT

I'd like to begin by thanking all our witnesses for participating in today's hearing on Building Decontamination.

Last week, several congressional committees held hearings on various aspects of our recent experience with bioterrorism. Concerns were raised about our ability to quickly and efficiently respond with appropriate assessment and decontamination protocols. A resonating conclusion was the need for more consistent information and more centralized leadership. One physician stated that we have been operating in an "informational vacuum." I agree and that is why we are here today.

Sadly, terrorism has become a fact of life. And although our law enforcement officials are working diligently to ensure our safety, we must take every measure to guarantee our preparedness. That means that we need to learn quickly from our current difficulties.

We are here today seeking knowledge in three areas. First, the coordination that goes into decontaminating a building. Second, the health effects of both cleanup technologies and residual contaminants. And finally, the various technologies available for remediating a building.

I believe that this hearing is critical as a forum in which we can all learn. After all, we are the test case, and the photos you see here today document an historical effort. Affected parties, such as the U.S. Postal Service, are awaiting a decontamination model to emerge out of EPA's current efforts to remediate the Hart Senate Office Building.

No prior attempt has ever been made to remediate a biologically contaminated building. In fact, 2½ years ago, the Working Group on Civilian Biodefense published a report which stated that such a decontamination effort would be extremely difficult. Well, leave it to Congress to expect to re-write science. EPA has been given a tremendous responsibility despite the lack of prior experience and systematic protocol.

Therefore, I am pleased that we have Governor Christine Todd Whitman, the EPA Administrator, with us today. I would like to thank her for her current efforts, and offer my assistance as we both learn about the response protocol necessary to effectively address acts of bioterrorism.

OPENING STATEMENT OF HON. BOB SMITH, U.S. SENATOR FROM THE STATE OF NEW HAMPSHIRE

Senator SMITH. Thank you, Mr. Chairman.

Good morning, Governor.

I wanted to make just a brief comment that Senator Inhofe has asked me to make but I share Senator Inhofe's concern because it seems every time there is an important Armed Services Committee meeting, there is an important EPW Committee meeting and many of us are on both of those committees. I know on at least four occasions Senator Inhofe has raised that point. There are four members who are on both committees. Again, Senator Inhofe asked me to raise it that we ought to try to have a little more coordination. I realize that is a two-way street but it really is a problem. I know there are some very important nominations that are taking place right now in the Armed Services Committee which means I am going to have to leave at some point before I wanted. I hope we can at least try to work to coordinate that a little better.

Senator JEFFORDS. We certainly will. We are not trying to be uncoordinated.

Senator SMITH. No, there is a chairman over there too. It takes two to coordinate. I understand.

Thank you for holding this hearing and Governor, it's good to see you again. You certainly had a baptism under fire with what has been going on. The anthrax matter has obviously been of grave concern to all of us. We are looking forward to hearing your remarks.

I have a prepared statement that I will enter for the record.

[The prepared statement of Senator Smith follows:]

STATEMENT OF BOB SMITH, U.S. SENATOR FROM THE STATE OF NEW HAMPSHIRE

Mr. Chairman, thank you for holding this hearing. Welcome to Governor Whitman—it is always good to see you before this committee. I also want to welcome all of the witnesses who are here today.

There is no question that these are difficult times. Beginning on September 11, this Nation has faced many of its worst nightmares—the attacks on the Pentagon and World Trade Center. And that was soon followed by the quiet horror of biological attacks.

Since September 18, several letters containing anthrax have terrorized this nation.

It has been devastating to our postal employees—and our Nation sends our deepest sympathies to those brave public servants who continue to do their duty in these very difficult times.

Those who are the innocent victims of this terror cover the spectrum, from a 7-month-old little boy, who was diagnosed with cutaneous anthrax—to a 94-year-old woman, who just recently died. We send our deepest condolences to families and friends of the victims of these cowardly attacks.

Of course, we here in the Senate, have also felt the sting of anthrax. Letters containing anthrax sent to the Senate have left many up here quite shaken. Most of us have been tested for anthrax exposure and many continue to take CIPRO as a precaution.

Twenty-eight Senate employees have tested positive for exposure, but fortunately, no infections.

While it has been a difficult time, we have been lucky that the difficulties have, thus far, only been inconveniences. These attacks have also left us with the dilemma of how to remediate the contamination of the numerous affected buildings.

There is an uneasiness with many who were in these buildings when the anthrax arrived and who will be going back into them when the cleanup is completed. We are all more than a little uneasy when dealing with so many unknowns—and I do not envy you, Governor Whitman, or anyone else involved in the testing and cleanup.

The Nation has many questions, concerns and fears—and the answers are not easy. It is certainly a daunting task. It is my hope that today, you and the other witnesses, will take this opportunity to address many of the questions that we all have.

There is much that has been done over the past few days and we are all certainly anxious to hear how the remedial activity of the Hart Senate Office Building went over the weekend. Hart is undoubtedly the testing ground for other anthrax cleanups, so it is important that the work is done in a deliberate manner. I hope that has been the case.

Governor, I do want to take this moment to thank you, Marianne Horinko and all of those at the Environmental Protection Agency for their tireless work since the events of September 11. It has been a new world for your Agency—an unquestionable challenge—and I commend you and the entire agency for your efforts.

Thank you all again for coming here today and I look forward to your testimony.

Senator SMITH. I ask unanimous consent to enter the opening statement of Senator Inhofe in the record as well.

Senator CARPER. Reserving the right to object.

[Laughter.]

[The prepared statement of Senator Inhofe follows:]

STATEMENT OF JAMES M. INHOFE, U.S. SENATOR FROM THE STATE OF OKLAHOMA

Mr. Chairman, thank you for holding this very important hearing. I, too, am very interested in hearing from Administrator Whitman on the challenges of, and technologies available for, remediating buildings contaminated by biological contaminants.

Specifically, I am interested in hearing from the Administrator on the following issues:

(1) Enhanced detection systems for chemical and biological agents; and

(2) EPA's compliance with all of their regulations during the remediation process.

Oklahoma is no stranger to terrorism. Until September 11th, Oklahoma had the unfortunate distinction of having been the victim of the worst terrorist act. As a result, what we as a Nation are doing right now; Oklahoma has been doing for a few years now. That is looking at how do we prevent and mitigate terrorist acts.

Since September 23, 1999, under the direction of former Army General and Chief of Staff Dennis Reimer, the Oklahoma City National Memorial Institute for the Prevention of Terrorism (MIPT) has been dedicated to preventing and reducing terrorism and mitigating its effects by conducting not only research into the social and political causes and effects of terrorism and but also the development of technologies to counter biological, nuclear and chemical weapons of mass destruction as well as cyber terrorism.

Originally incorporated as a non-profit corporation in Oklahoma and recognized as a charitable organization by the Internal Revenue Service, MIPT grew out of the desire of the survivors and families of the Murrah Federal Building bombing of April 19, 1995 to have a living memorial. As Oklahomans, we intend to honor that desire by doing what we can to try to prevent other cities from living through what Oklahoma City had to live through—and what New York and Virginia are living through now.

MIPT has a special obligation to first responders—police officers, firefighters, emergency medical technicians and all of the others who are first on the scene in the aftermath of terrorist activity. Therefore, they also sponsor research to discover equipment, training and procedures that might assist them in preventing terrorism and responding to it. While MIPT has a special obligation to first responders, they are prepared to engage in any activities that will help them fulfill our mandate.

For example, unfortunately, today's anthrax vaccine is not appropriate for protecting the general public, so there is a critical need to develop new therapies that could be quickly administered following a bioterrorist attack. Therefore, just recently, MIPT and the Oklahoma University Health Sciences Center started a 3-year, $2.48 million research effort to develop new drugs which will lessen the threat from anthrax. Specifically, this research seeks to develop new medications that block the lethal toxins produced by anthrax bacteria. These medications could be much more effective than the current vaccine since they would target toxin activity after the initial anthrax infection.

With this tool that may counter anthrax more effectively than the current vaccine, the United States may be better positioned to deter terrorists from considering this type of weapon in the future.

This project is one of 10 counterterrorism projects that MIPT is currently pursuing. Other projects include better protective clothing for those working in hazardous environments, enhanced detection systems for chemical and biological agents, a study of communications surrounding terrorist episodes, a study of the psychological impact of terrorism, defense of communications systems, and data bases on terrorism and counterterrorism equipment.

MIPT currently funds projects all over the country, including California, Florida, Missouri, Rhode Island, and Virginia. MIPT has also received over 250 proposals for its next round of research projects. A decision on which of these projects to fund will be made in the course of the next 90 days. As we move forward with preventing and mitigating terrorism, I would urge my colleagues to work with MIPT. Perhaps, the testimony, which we hear today, can help provide some ideas to further MIPT's critical work.

OPENING STATEMENT OF HON. THOMAS R. CARPER, U.S. SENATOR FROM THE STATE OF DELAWARE

Senator CARPER. I will not object if the chairman would let me make a short statement.

Governor, welcome.

I am one of nine Democrats on this committee whose offices are in the Hart Building. We have been without an office for about 1½ months. But for the grace and hospitality of Senator Biden—his staff, we would be out on a truck loading dock somewhere. As it turns out, he has been good enough to share his conference room with us. It is tight quarters but at least they are quarters and we are grateful for that kindness. It has given us a close knitness as a congressional delegation, a Senate delegation, as you might imagine, all being huddled there together.

I am sort of torn on the issue of Hart. On the one hand, we want to get back, not desperately but expeditiously, into our quarters to be able to do our jobs better. By the same token, we want to make sure we are all safe. It is a tough balancing act and we are real grateful for your help and that of your agency, first of all, to make sure when we do go back, we are safe and second, we can get back in there as quickly as advisable.

That being said, Mr. Chairman, I yield my time. I am going to go preside at 10 o'clock and won't be able to stay for part of this hearing but I look forward to your testimony, Governor. Welcome.

Senator JEFFORDS. Thank you.

Please proceed.

STATEMENT OF HON. CHRISTINE TODD WHITMAN, ADMINISTRATOR, ENVIRONMENTAL PROTECTION AGENCY

Administrator WHITMAN. Thank you very much, and with your permission, I have a longer statement that I would like to submit for the record.

Senator JEFFORDS. It will be.

Administrator WHITMAN. Thank you.

Since the events of September 11, the Environmental Protection Agency has seen its longstanding mission to protect human health and the environment take on new meaning and a renewed sense of urgency. Since the discovery of anthrax in various public and private buildings, EPA has been operating under the authority or response structure that we have long used in addressing Superfund contaminations which gives us a responsibility for cleaning up contaminated sites to protect human health and the environment. The manner in which we have proceeded also follows the general provisions of the Presidential Decision Directive No. 62 signed by President Clinton in 1998. That PDD gives the Environmental Protection Agency responsibility for cleaning up buildings and other sites contaminated by chemical or biological agents as a result of a terrorist act. Nevertheless, as we have moved forward to address our responsibilities in the anthrax contamination, we have found the structures and responsibilities outlined under PDD No. 62 have not been as clearly articulated as we would like because everyone involved has been focused on getting results and we have not allowed the discussion of a process to halt the progress. Even though it seems like a long time, that has not been the case. The lack of specific clarity in the 1998 Presidential Decision Directive has made things more complicated at times than they needed to have been.

I raise this issue not to be critical in any way but rather to highlight the fact that plans put on paper almost always have to be adjusted to the realities of events and that is why we are working very closely with Governor Ridge and the Office of Homeland Security to bring greater clarity to the roles of the various Federal departments and agencies they should play in responding to biological attacks and how we should best relate to our State and local partners.

In responding to the current biological attacks, EPA's role at the site generally began after the Center for Disease Control and Prevention determined the presence of a biological contaminant that posed an unacceptable risk to human health. We have also worked with CDC to advise the incident commanders about the extent to which the building must be cleaned to make it safe. Then once a building has been decontaminated, the incident commander has responsibility for determining whether a building is safe for reoccupancy.

The sites themselves are under the control of the incident commander. Usually someone from the local response team, the EPA, CDC and the other Federal agencies work with that incident commander providing expertise and advice and performing such work as testing and cleanup at times.

With respect to the cleanup of those places found to be contaminated by anthrax, several different approaches have been taken. The Postal Service, for example, has hired a qualified contractor to perform cleanup at their facilities, as did several media organizations. In those cases, the Environmental Protection Agency provided technical assistance to those actually doing the cleanup work.

Here on Capitol Hill, we have been asked by the Senate Sergeant at Arms, who is serving as your incident commander, to undertake the cleanup of the Hart Senate Office Building, just as we were asked by the Clerk of the House to clean up the contaminated locations on the House side.

As you know, the cleanup of the Hart Building poses by far the largest and most extensive cleanup challenge of anthrax ever undertaken in a building. To meet this unprecedented situation, our cleanup experts have been drawing on their years of experience and expertise, on the talents of scientists and industry and academia and the knowledge available from our other Federal partners to devise the right plan for the Hart Building. This expertise and experience has served us well to date, leading to the successful cleanup of many post offices and other buildings and this past weekend to the fumigation of parts of the Hart Building. As our knowledge increases, our ability to successfully address anthrax contamination will continue to improve. We are quite literally writing the book as we go along. I am proud of the work the agency has done in identifying methods to clean up anthrax in situations never before envisioned by planners or prognosticators.

It is not an exaggeration to say that EPA and our partners have done more in the past 6 weeks to advance the knowledge about the science and technology of anthrax detection and cleanup than in the previous six decades. As we seek to apply the lessons we are learning from all the decontamination efforts from the simplest to the most complex, one thing has become very clear and that is that one-size-does-not-fit-all when we are facing this kind of challenge.

Each event has to be thoroughly analyzed as a separate case before we can propose an effective solution.

For example, cleaning a facility that contains rugged heavy equipment can be accomplished generally using foams or liquids, methods the contents of that building can stand up to. A building that contains a lot of paper, office furniture, electronic equipment, that must be cleaned up using a different method so as not to damage those things the way a liquid or a foam would damage them.

Other factors such as the amount of contamination found, the ways and the extent to which it can be dispersed through the building, the nature of the surrounding areas and the ways in which a building is used all require different considerations before proceeding with decontamination. That is why it has required more time to address the Hart Building than any of us would have liked. This has been a highly complex challenge but we believe the time we have taken has been well spent because we have taken the time to do it right and we have advanced our knowledge in the best way to undertake these efforts.

While we are all hopeful we are never going to have to use this knowledge again, we have to proceed as if we might. That is why there are two specific things I would like to ask the committee to consider for the future. The first concerns indemnifying the contractors EPA hires to perform the actual cleanup. We spent a great deal of time in recent days leading up to the fumigation in the Hart Building last weekend to provide the contractors we were hiring to perform that decontamination with sufficient protection and liability should something unexpected occur during the cleanup. After a lot of work, we were able to get the issue sorted out but it took much more time than it should have.

EPA's current indemnification authority under CERCLA is not adequate to meet the needs resulting from acts of terrorism. That is why it would be helpful if for the future, EPA's indemnification authority could be extended to meet the response to domestic acts of terrorism.

The second issue where I would like to ask for your help concerns our ability to recover costs resulting from cleanup. Currently, EPA can recover costs when performing the cleanup of hazardous substance. The authority, however, does not extend to biological agents or various other pollutants that could be used in a terrorist attack. Giving EPA the ability to recover costs in those instances would just remove one more issue that slows us down a little bit, would allow us to focus a little more quickly in getting things operating faster.

I should point out that this has not been an issue with the Hart Building. The Senate has agreed to pay from the beginning and we are very thankful for that and appreciate that.

I would be happy to take any questions you might have on any of these issues.

Senator JEFFORDS. Thank you. We are all sitting here interested in this. My office is in the Hart Building. Yours is too?

Senator SMITH. No, mine isn't. It is in Dirksen, I am pleased to say.

Senator JEFFORDS. How did EPA identify chlorine dioxide as the best remediation technology?

Administrator WHITMAN. What we did is we were in close contact with the CDC, with the military, we talked to the private sector and we had a trailer set up at the Brentwood facility where we used the different methods of decontamination. First, understand there was no licensed product to decontaminate anthrax in a building. We have had to issue some emergency clearances to allow us to use the chlorine dioxide liquid and the envirofoam that are the two agents you put on hand application and then the chloride dioxide fumigant, the spray.

We looked at what would have the greatest impact, what actually resulted in killing of the spores in the DNA and the spores and the collateral damage that would be done to the rooms that were being decontaminated or anything that might be in those rooms. Based on that, we put together a proposal we submitted to the incident commander with recommendations as to how to proceed.

Senator JEFFORDS. What remediation technology beside chlorine dioxide did EPA test?

Administrator WHITMAN. We did test the chlorine dioxide liquid as well as the spray, the envirofoam and a number of other fumigants, disinfectant bleach solutions, chlorine dioxide liquid, fumigants, chlorine dioxide gas, ethylene oxide, vaporized hydrogen peroxide, the HEPA vacuuming in almost every suite that has shown contamination. Every suite that has shown contamination will be HEPA vacuumed. We have looked at radiation and destructive disposal which includes incineration and autoclaving. So we have been pretty thorough in our review of what is available.

We are, I will tell you, receiving daily more products, more technologies that have been given to us as new methods to use. It is wonderful what happens in this country when there is a problem, how people rise to meet it, and we are in the process of analyzing all those to see their effectiveness and fast tracking them to try to move through the process as quickly as possible.

Senator JEFFORDS. Did EPA ever consider remediation technologies that would take the risk of secondary aerosolation into account?

Administrator WHITMAN. We looked at everything we thought was appropriate, where we thought we could not only achieve the goal of zero contaminant left in the building but also ensure there was going to be no residual effect from whatever methodology was used, what we felt we could guarantee and show was completely out of the building and the buildings were therefore safe. Actually the CDC will be the final determinator of when it is appropriate for humans, what they consider to be safe for everyone. Then the incident commander will determine when exactly they can move back in.

Senator JEFFORDS. Now that you have fumigated Senator Daschle's office, what is the next step? May you have to fumigate again and are you considering other technologies if chlorine dioxide does not work?

Administrator WHITMAN. Right now we are waiting to get the results from the suite. We should have those by the end of this week. We have sent them to the labs, the strips and we put in dishes to do additional testing to make sure both the chlorine dioxide gas is out of the suite and also that we have achieved the results of zero trace of living anthrax in that suite. We should have that by the end of the week. By determining what that tells us, we will know what the next steps would be. There are a number of different options that we could do. We could go back in and fumigate again, we could do the wipe using the envirofoam or the chlorine dioxide liquid. Those are all options we would be discussing with the Sergeant at Arms who is the incident commander.

Senator JEFFORDS. What lessons has EPA learned about responding swiftly to an emergency situation in which defined procedures may not necessarily be in place while at the same time allowing for public input?

Administrator WHITMAN. Actually, we have learned that overall it works very well, that as I indicated to you the Presidential Decision Directive that gives us the responsibility to respond in cases of a biological attack is not perhaps as clear when you start operating as one would like it to be, that basically when you put aside all the turf battles, and those went aside very quickly, we found it is working pretty well. We need to enhance and improve the monitoring and health safety for on-scene coordinators. That is an issue of concern. As you may remember, from the beginning there was discussion of whether they needed to put on the breathing apparatus right at the beginning; there was some discussion and concern about what was the appropriate protocol for protection as it was for you and your staffs of Cipro. We need current medical monitoring that needs to include a pre-response screening. We need to have our people ready. It is something we are looking at in our labs as well.

It is interesting that if you are getting people in the labs to do the testing on something like anthrax, there is a real concern about allowing someone who hasn't been either vaccinated or isn't on something like Cipro to actually do the testing. Some of those take some time. You buildup immunities over time and we need to make sure that we know what the protocols are and that we have the appropriate people who are appropriately protected responding.

Those are some of the areas in which we would like to see more attention as we move through this.

Senator JEFFORDS. What is EPA's protocol for alerting the public to current or ongoing emergencies, whether that be in a public building or chemical facility?

Administrator WHITMAN. The response team is the overall group that responds in any instance. The Center for Disease Control is the agency that has the primary responsibility for determining threats to human health. They will test, we will test with them, and if we see any indication there is something that poses a threat to human health, that's when we would respond with the CDC and alert the public to the extent that is necessary to ensure their protection. We would do it under the guidance of the CDC.

Senator JEFFORDS. Thank you.

Senator Smith.

Senator SMITH. Governor, you mentioned a moment ago one of the options was vaporized hydrogen peroxide. Could you or anyone on your staff indicate to me why that was ruled out? There have been some who argued that might have been the safer procedure and do the job just as well. I am not trying to second guess you; I am just trying to clarify it for my own mind here.

Administrator WHITMAN. I could certainly ask one of the experts to come up and give you a more thorough response or we could do it for the record or afterwards, but let me say that in all instances, we looked at a number of different options and looked at both the immediate ability to kill the anthrax spores and the residual impact it might have, and determined that the chlorine dioxide spray we are using in the Daschle suite is one that if you maintain the right humidity level and the right temperature, which we were able to do over the weekend, proves a very effective agent and at the same time is one we can assure we can get out of the suites, out of the air systems and you have to remember is basically something that is used in everyday products. Chlorine dioxide is used in water, used to spray on vegetables, so it is something that at the appropriate levels has no adverse human health effects.

Senator SMITH. Was there any contamination found in the Hart garage?

Administrator WHITMAN. Not to my knowledge.

Senator SMITH. You did test there? Is there a different vent system there?

Administrator WHITMAN. Yes, we tested the garage, and no, we did not find any contamination.

Senator SMITH. Because cars were allowed to park there throughout the entire process, even before you tested, it seemed a little odd to me that would happen. I am assuming you haven't ruled out anthrax on any particular floor or section of the Hart Building when you proceeded. Is the Hart garage under a different vent system? Is that the issue?

Administrator WHITMAN. We have tested now every suite in the Hart Building and tested the garage. A lot of this appears to be collateral contamination from the original letter. So it isn't necessarily so that it would get into the entire system, into the air vent system or be picked up by every employee, staffer or Senator who had been exposed to it. That somewhat limited the cross contamination but we did not eliminate anything when we looked at the building and we have tested all the areas.

Senator SMITH. Was the anthrax in the ventilation system of the Hart Building?

Administrator WHITMAN. There was a hit on the Senator Daschle suite and that was one of the systems that we worked on immediately.

Senator SMITH. No other place in the vents other than right there?

Administrator WHITMAN. No, not in the vents. The garage is on a different HVAC system.

Senator SMITH. Have there been any other offices where you have found more anthrax since the initial hit at Senator Daschle's office?

Administrator WHITMAN. There are 11 offices where we have found traces of anthrax and those are all being treated in ways appropriate to the level of contamination. The Daschle suite obviously had the most because that is where the letter was opened, so there was the most contamination there which was why we thought it appropriate to use the gas.

As you may remember, originally, we thought we could do the entire building the same way, just to make extra sure even though most of or large parts of the building had no contamination at all, but the engineering of that proved to be too difficult to ensure the right outcome. So for the Daschle suite, we used the spray. For the other suites, we are going in with the envirofoam and the chlorine dioxide liquid and wiping anyplace that we found contamination, HEPA vacuuming the entire suite first, then using either the foam or the liquid on all areas where we got any hits from the testing for anthrax.

Senator SMITH. As far as you know, and maybe you don't know, all of the anthrax found in the Hart Building came from the Daschle letter?

Administrator WHITMAN. As far as we know, yes.

Senator SMITH. Just in the overwhelming task you have before you, a giant building, several floors, just in a moment and not a long answer, how did you go about starting this? What did you have to do? Did you go floor by floor, hall by hall trying to secure? Give me a process of how you go about starting this and working through this?

Administrator WHITMAN. Initially, NIOSH did the preliminary testing. They actually followed what they thought to be the path of the letter itself. We came in when they needed a little extra help and did some additional testing. We followed the trail of the mail, where mail would be handled in an office, where it would be sorted in an office, and that is when we picked up some additional contamination. We then decided to test the entire building, every suite and in the suites where we found any contamination, we tested every room. We were determined to be as thorough as we could. It was a protocol we worked out with the CDC, with our other Federal partners and the private sector and presented to the incident commander. They had the final approval for what they felt was the appropriate protocol.

Senator SMITH. Thank you. We appreciate the job you are trying to do. It is a tough job.

Administrator WHITMAN. We are working on it.

Senator JEFFORDS. Senator Corzine.

OPENING STATEMENT OF HON. JON S. CORZINE, U.S. SENATOR FROM THE STATE OF NEW JERSEY

Senator CORZINE. Thank you, Senator Jeffords. I appreciate your holding this hearing, Mr. Chairman. It is always good to be with Administrator Whitman and renew ties. Unfortunately, some of those ties come because we have more than a few of these events seemingly originating in New Jersey.

One of my concerns is are we applying similar standards of cleanup to New Jersey post offices and other post offices that we are now establishing with respect to the Hart Building? Are these protocols going to be administered by the EPA on a consistent basis across the country if this were to have greater legs than what we see today?

Administrator WHITMAN. Although I am somewhat reluctant to say this to Senators, they are guinea pigs. Neither the Hamilton Post Office, nor the Brentwood facility have been decontaminated yet and they are actually watching to see how the decontamination, the work on the Daschle suite has gone. That will determine how they are going to proceed. The post office is the primary responder there. They will make the determination. We act as consultants. We will provide them with advice, anything they want but they are the ones who are going to hire a contractor to do the decontamination.

Senator CORZINE. So we won't necessarily have a consistent pattern of cleanup on how we address this across the country?

Administrator WHITMAN. No, it depends on the incident commander. Each site has its own incident commander, but the other thing that is very important to remember is it really does matter the type of contamination, where you are finding it and the type of room. For instance, the post offices in general, those rooms have heavy equipment, we are able to do a better job with a wipedown and using either a foam or a wipedown. As I mentioned earlier, we are looking at a number of different technologies, so it will depend on the type of contamination, how much and where we are finding it, and what we recommend.

Senator CORZINE. After we have gone through this process in several instances, do you think we ought to get to a standard, not a standardized approach because it happens a different kinds of facilities, but should there be a standard protocol with regard to who is responsible for administrating this, who is responsible for controlling the costs, using the application that seems most appropriate? It seems the building of expertise, at least from my perspective, would argue that we would like to see the same people doing this, not reinventing wheels on a regular basis.

Administrator WHITMAN. Actually, there is such a process in place in the National Response Team. There are the same agencies that are involved in that. The problem you have with many of these is you don't want to usurp the authority of the State or local responders. That is the way emergency response has gone. We do need to take a look with this kind of biological terrorism to see whether the Federal Government should come in more quickly but the CDC is called in almost immediately on most of these kinds of instances. Also, we need to remember they are crime scenes, so the FBI is there and they control the site initially once there is a determination you have had an anthrax or biological contamination.

We are the ones tasked with being the Federal responders on biological decontamination when the Federal Government is called in. We also have the ability if we think a local entity or private entity, whoever is responsible for the cleanup, if we don't think they have the capability to do it, if we don't think they have the technical expertise or the willingness to do it appropriate, we could step in. That would require at this point getting a court order in order to do that, which is not something I don't think any of us is particularly anxious to do. Thus far, it has not been a problem but there is a procedure in place for how this would occur.

Senator CORZINE. The request you made at the end of your testimony with regard to indemnification and recovery costs, would those be amendments to the Superfund Act? Is that where you would most appropriately see those?

Administrator WHITMAN. That could be one way to do it but we are using CERCLA because that is the money we have for emergency response, but it would not be inappropriate to have a separate indemnification. It would be under CERCLA, under Superfund.

Senator CORZINE. Finally, several weeks ago we had a hearing on chemical security. You and I had some discussions about preparation and timing in which one might want to comment on some of the elements of that. I was curious whether you had a chance to move forward with the review and whether you would be available on that subject?

Administrator WHITMAN. We certainly appreciate the goals of the legislation and we have submitted our comments to OMB, so we are getting close to an Administration position which would allow us to comment.

Senator CORZINE. I would suspect after the first of the year, we would like to revisit that. It continues to be a concern that I hear quite frequently from constituents and others.

Thank you.

Senator JEFFORDS. Senator Voinovich.

OPENING STATEMENT OF HON. GEORGE V. VOINOVICH, U.S. SENATOR FROM THE STATE OF OHIO

Senator VOINOVICH. Thank you for holding this hearing this morning, Mr. Chairman, and thank you for being here, Administrator Whitman.

There is a lot of uncertainty in the country today and a lot of anxiety. How soon did you get involved with the Hart Building?

Administrator WHITMAN. We were asked to get involved with the Hart Building as an observer providing just some expertise right from the beginning and we started to do the testing toward the end of October. We got involved in the actual testing. We were up there providing some advice from the beginning.

Senator VOINOVICH. There was a great deal of confusion and I am still unhappy about the information we got, particularly when I told my staff members not to get their nose swabbed, it wasn't necessary and that they were fine, and we then woke up and read in the paper that things were different. I think that kind of information, what happened at the post office, two postal workers are dead today because it could have been handled differently. We got the mail and nobody was concerned about the post office where the mail was run through.

I am interested in the Hart Building and frustrated about it like a lot of others. I would say, one, I would hope we are going to have some witnesses here and I know it is a learning experience to a degree for your agency, but I would be interested after it is over that good records are kept about whether this works or doesn't work and also that there are others out there with products on the market, and I would hope you would give them an opportunity also to be tested so there is some kind of good housekeeping, whatever it is you can give to people to say yes, this works because hopefully we are not going to have a repeat of this but if we do, I think people ought to know the next time around, we will be in a lot better shape to respond than we have been able to do thus far.

The area I am really concerned about is the post office. Are you involved with the post office?

Administrator WHITMAN. We are just providing advice to the post office. They are the primary responders and are taking responsibility for that. I can give you a timeline if you would like for the Hart Building too. I would be happy to submit that for the record if you want, of when we came in, what CDC did, and they established the human health aspect. That is not something we do. Our primary responsibility is decontamination. We don't even usually do the testing.

Senator VOINOVICH. You are not involved in the post office except as a consultant?

Administrator WHITMAN. Except in providing them sage counsel and advice in everything we have seen. That is why I indicated they are waiting to watch, to see how successful the decontamination of the Hart Building is, particularly the Daschle suite activities undertaken last weekend. Predicated on that, I think they will move forward with a similar type of decontamination.

Senator VOINOVICH. I would hope you would share with your colleagues in the Administration how important the post office situation is because there is much anxiety in the country today. I have constituents that say they don't want us to respond to any letters they get from Washington.

Administrator WHITMAN. That could be something they wanted all along, Senator.

Senator VOINOVICH. I have a daughter-in-law that says, Dad, I don't want any mail from Washington anymore. I think that is something that is adding to the fear. There is a lot of fear in our country today. I think those of us in Washington really don't get it. It is pretty severe. I think it is really important that in the process of doing the Hart and some of these other things, that we get some good information out there to the American people about the fact that we are a lot better prepared to deal with any new things and certain areas they are concerned about, they ought not to be concerned about them.

Administrator WHITMAN. We will be providing an administrative record that is a public document at the 60-day time period that will indicate everything we have done, what we have looked at and will be available to anyone to see. We are working closely and as you indicate, there is a lot of concern in the country. The Center for Disease Control is the one setting the standards and making the determination as to what is safe for humans. We follow their lead and provide decontamination to reach the goals they set for that decontamination process. We have been working very closely together.

On your other issue of alternative methods of decontamination, we have about 30 new products that have been submitted to the agency since the anthrax letters first turned up in early October. We are fast tracking that to the degree possible to ensure they do achieve what they say they are going to achieve, that they are safe to be used in these instances, and try to get them out there. I do believe along with you that by the time we are finished with this, we will be very much more able to respond, will have many more options at our disposal and a wider range of options, so we will be able to get things done faster. We are literally writing the book as we go along right now.

Senator VOINOVICH. Mr. Chairman, if possible, I would like us to get a request out to the Administration to Governor Ridge about just how is the Administration coordinating all this? Who is responsible for what? Who is the quarterback? You said you weren't involved except as a consultant with the post office. Who actually is the one in charge? Is it CDC that is calling the shots at the post office? Who is in charge?

Senator JEFFORDS. I will be talking to Governor Ridge. We are arranging a meeting now, so I will extend those questions to him.

Administrator WHITMAN. Let me say the incident commander in all those instances is the final arbiter of what advice they take, what is safe. They take the advice from EPA, CDC, the Army, from a lot of different people but it is up to them at the end of the day to make the determination. In the case of the post office, the post office is the incident commander. Here, it is the Sergeant at Arms.

Senator VOINOVICH. Thank you.

Senator JEFFORDS. Senator Clinton.

OPENING STATEMENT OF HON. HILLARY RODHAM CLINTON, U.S. SENATOR FROM THE STATE OF NEW YORK

Senator CLINTON. Thank you, Mr. Chairman.

I want to associate myself with the questions and comments of both Senator Corzine and Senator Voinovich on the matter concerning the protocol for use and the responsibility for overseeing the response to anthrax. I appreciate very much your describing to us in your written testimony, as well as here, what you are doing.

I think we do have a lot of questions we need to answer. There are questions everyone is trying to answer. There isn't any standard response out there yet, but that is what we are intend upon learning and trying to create. On October 26, I actually wrote to the Postmaster General and the Centers for Disease Control asking we begin the process of trying to adopt a standard protocol. We look forward to working with EPA and every other agency that is affected.

The impact on my colleagues, including EPW staff because of the discovery of anthrax and its effects in Hart have been extremely difficult. We know it is even more so for our citizens who have been working in post offices. I want to be sure that whatever we do, we do for everyone, that we don't have some special treatment for Members of Congress as opposed to people working in postal offices or any other facility.

One question I would have specifically out of your written testimony is the agency has apparently approved two pesticides for treating anthrax spores under emergency exemption provisions of existing pesticide laws. I would like written response and more explanation of what that means. How does it work? I believe the emergency approval was to permit actions to be taken in postal facilities. I, along with my colleagues, hope we are getting good information about comparing what we have done in the Hart Building with comparing the use of the aqueous solution of chlorine dioxide and a foam of some kind, just so we know what we are learning and what we are doing.

I would also be interested in how you are currently paying for the work you are undertaking since I believe you asked in your testimony for, among other things, the ability to recover cleanup costs. How is that being paid for now?

Administrator WHITMAN. Right now, we are working under CERCLA. We have spent about $7.5 million to date nationwide. We expect that to get significantly higher before we are finished with it. The Senate has agreed to pay for the decontamination here. That is an area where we have some concerns. There is not a responsible party here in the traditional sense because this was done by a third party unknown to everyone who has been impacted by it thus far. So we are using our CERCLA moneys at the moment to do that.

Senator CLINTON. I would just note in Senator Byrd's Homeland Defense package we will be considering this week, there might very well be some funds that could be used to reimburse EPA for the work it is doing right now.

On a different environment and public health issue, I would like your response about Ground Zero and the area surrounding the World Trade Center. As we all are unfortunately aware, the fires are still burning. If you saw the paper today, there is going to be an effort to remove the gas that was used to cool the towers, something that is extraordinarily difficult to undertake. Certainly, I hear from a lot of people complaining about the air quality asking questions about the environmental and public health concerns, whether it is asbestos, dust or high levels of benzene or thousands of gallons of PCB laden oil being released.

I sent a letter on October 26 expressing my concerns on this issue and requesting a meeting and I was scheduled to meet with Assistant Administrator Maryanne Horinko, who I am pleased to see is here. Unfortunately, she had to cancel that meeting last week which I hope will be rescheduled for this week. I look forward to meeting with you because I think it is imperative that we make every effort to provide the best possible information to people about what is happening in the air quality testing and that the public can make decisions concerning their own well being.

One of the big issues we are facing right now is whether to bring elementary school children back down to Ground Zero to be in their schools again. I don't know what to tell parents. I don't know what I would do if I were the mother of an elementary school child. We did reopen Stuyvesant High School. The kids are going to school, the air is being tested but there is a lot of what we are now calling World Trade Center cough, respiratory asthma problems and no one is quite sure whether we should go with the younger children.

Mr. Chairman, I think this would be a good matter to hold a hearing about when we get back after the holidays to try to figure out what we should be doing and how we can provide good information.

Finally, I couldn't have the Administrator appear before the committee without asking about the Hudson River cleanup. I would like, if I could, a brief update on the record decision and when you expect that to be sent to the State. I know the Governor and many of us have been working very hard behind the scenes to make sure this was carried out the way we believe it should be. We are opposed to the agency including performance standards. I hope you are going to be able to give me some news today about where we stand in that process.

Senator JEFFORDS. I am sorry, your time has expired.

[Laughter.]

Senator JEFFORDS. Please proceed.

Administrator WHITMAN. We hope to have that out very shortly and I mean very shortly.

Senator CLINTON. Is that within the time period that I am no longer able to—very surely by the end of this week?

Administrator WHITMAN. I hope so.

Senator JEFFORDS. Thank you very much. We deeply appreciate your help and we will be continuing to communicate with you and try to get a better understanding of where we are and where we are going.

A question I would sort of like to end with is when does zero arrive in determination of a risk in these situations? Those of us involved with farmers know that anthrax is in the fields and yet we seem to be looking for one spore to close things down. What kind of guidance do we have as to when it is a sufficient problem to take action?

Administrator WHITMAN. Senator, there is no background level that has been determined of anthrax in an urban setting or in buildings and that is why at the moment the goal advocated by the CDC is zero. That is what we are operating toward. We have cleaned up three offices thus far in the Hart Building and they are showing no sign of contamination. They have been successfully remediated. We have remediated a couple of post offices—we have overseen the remediation of a couple of post offices in Florida where they did call us in, Brentwood they have not, to take a more active role and those have been cleaned and show no background. So we are comfortable that we will be able to reach that understanding.

As you say, anthrax is a naturally occurring agent, not in the kind of form and milled to the fineness that the anthrax in the letter to Senator Daschle was milled. That was very refined. It was able to get through the envelope itself without having been opened. They are finding that in fact it could get through the paper itself. So this is a different type of contamination of anthrax, but the CDC is looking at and will make the determination of what is safe, whether there is a safe level of anthrax for human exposure. Until that time, until they make that kind of determination, we will be going for the goal of zero anthrax and thus far, we have seen that.

Senator JEFFORDS. You mentioned that the Presidential Decision Directive No. 62 needs improvement. What changes do you think may be necessary to strengthen the Presidential Decision Directive No. 62 and do you anticipate recommending these changes to the President?

Administrator WHITMAN. We are working very closely with the Office of Homeland Security through lessons learned on this whole issue, analyzing where we think there could be better coordination, where we feel there is more need for focus. For instance, as I indicated, what do we need to provide our responders. We have had people in the Hart Building from the very beginning. Initially they didn't have all of the protective gear that they now are wearing, although we took extra steps right from the beginning but we need to come to a better understanding of how those determinations are made and coordinate that more closely.

Senator JEFFORDS. Senator Voinovich.

Senator VOINOVICH. I would like the statement I have prepared inserted in the record and I have no further questions.

[The prepared statement of Senator Voinovich follows:]

STATEMENT OF SENATOR GEORGE V. VOINOVICH, U.S. SENATOR FROM THE STATE OF OHIO

Good morning, Mr. Chairman. I would like to thank you for holding this important and very timely hearing into the remediation efforts of biological contamination of buildings.

Since the first time anthrax was used as a weapon of terror, the one thing that the American people want is information, such as what to look for, how to avoid exposure and how best to deal with contaminated mail. Unfortunately, the availability of this kind of information has been spotty, at best. Since October 15, I have personally been very frustrated with the quality and reliability of the information regarding the anthrax contamination here in Washington, including the clean-up efforts.

Like many of my colleagues, I was originally told that my staff and I were safe in our office on the 3rd floor of the Hart Building, that we had nothing to worry about. As news stories trickled out that the spores in the letter mailed to Senator Daschle were of a pure and highly potent variety, I was still informed that my staff and I had nothing to worry about and that we should go about our daily routine, even though several members of my own staff who work for the Senate Governmental Affairs Committee on the 6th floor of the Hart Building were put on a 60-day regimen of Cipro. Two days later, I read in the *Washington Post* that the Capitol Attending Physician recommended that anyone who was in the Hart building, even for just a short while, over the previous 2 days should be tested for anthrax exposure: Senators, staff, constituents, couriers—everyone.

While we here in the Senate were expressing our concern over and reacting to our specific anthrax situation, workers at the Brentwood Mail Facility—where the letters targeting Capitol Hill were processed—were evidently unaware that letters laced with anthrax had passed through their building. Even more frustrating to me and to many others is the apparent fact that the safety and well-being of postal workers who handled the contaminated letters was given nowhere near the same consideration as was given to Senators, staff and visitors. Because of this inconsistency, two postal workers needlessly lost their lives.

Last month, I visited two post offices in Ohio to meet with the workers there to reassure them that people here in Washington are concerned about their safety. I also let them know that the Federal Government is doing everything possible to guarantee that they are not put in danger simply by doing their jobs.

I am also concerned about the more than 50,000 Ohio residents who have written my office in the last 6 weeks. Like my colleagues, I pride myself on being able to respond to my constituents in as timely a fashion as possible. However, given the disruption in the Senate's mail service, a vital communications link between my constituents and I has temporarily disappeared.

In fact, it was only just yesterday that we started to get a trickle of mail. God only knows how we are going to handle the mail in our temporary quarters when it starts coming in a full force.

Mr. Chairman, the thrust of the anthrax contamination has directly impacted several of our colleagues here in the Senate as well as their staff, but it has had the ancillary effect of forcing 50 Senators and hundreds of staff members from the Hart Building. Over the last 6 weeks, a variety of public officials from Capitol Hill and Federal agencies debated the best course of remediation for the Hart Building until one was chosen, and it will likely be several more weeks until we are able to reenter our offices.

We can do better, and I am looking forward to hearing from our witnesses as to how we can do so. One such witness is Mr. Les Vinney, president and CEO of an Ohio-based company, Steris Corporation Mr. Vinney has a significant amount of experience handling dangerous biological and chemical contaminants.

I would be interested in hearing what Administrator Whitman has to say regarding protocols that are being used in the Hart Building clean-up effort, particularly since the EPA is typically the regulating entity over a cleanup project such as this. Since the Agency has come under criticism for its handling of this cleanup, does this experience give you a new perspective on the difficulties that companies and other groups the EPA regulates face on similar projects?

The main thing I want to know is: what lessons have we learned from this experience and can we assure the American people that we have our act together? In addition, I have a special interest in ensuring the safety of the mail and particularly, our U.S. Postal Service employees. Since future anthrax attacks remain possible, these men and women are truly on the front lines and deserve our support.

Mr. Chairman, thank you again for calling today's hearing.

Senator JEFFORDS. Senator Corzine.

Senator CORZINE. Administrator Whitman, did I hear you say that you were invited or requested in the Florida postal situation and not in others?

Administrator WHITMAN. We are assuming the role we were asked to assume in Florida. There were three postal buildings in Florida where we were asked to actively oversee land be in the building as they decontaminated. They have moved forward with decontamination, they hired contractors, moved forward with decontamination. We have done the subsequent testing.

Senator CORZINE. Are you supervising that, the decontamination in Florida?

Administrator WHITMAN. We are technical consultants and coordinators.

Senator CORZINE. The whole point I am driving at is the same one we have heard several times here. One time you are in, sometimes you are not. I don't think people are going to take great confidence in knowing that there are broadly different approaches to this. It is not particularly fair to your staff, nor certainly not fair to the public at large. If one thought that the Hart project was going to get the Triple A fashion treatment and others not, I think there will be serious misgivings in the public and rightly so.

Administrator WHITMAN. There is no difference in the approach taken, be it the thoroughness of the approach taken. Anything that where we are consulting and acting as technical advisers, the standards are exactly the same for everybody.

Senator CORZINE. If you are acting as technical advisers.

Administrator WHITMAN. Yes.

Senator CORZINE. That is a big difference across the way. This is not like you are going to have 100,000 incidents. We hope it is not going to be the kind of thing where you have or we hope we don't have to have a lot of practice so where the best practices reside, I think the public has a reasonable right to expect that we apply them on a consistent basis.

Again, I want to underscore multiple times this idea of trying to get to best practices and making sure we have a coherent and consistent protocol on how we deal with this.

Senator JEFFORDS. Senator Clinton.

Senator CLINTON. I just want to be absolutely clear that EPA has not been asked for help with the anthrax-contaminated buildings in New York?

Administrator WHITMAN. No, the NBC buildings were done by private contractors.

Senator CLINTON. And the postal facilities?

Administrator WHITMAN. In New Jersey, they have not done anything at the Hamilton facility. That remains closed to date. We have an on-scene coordinator there working with them.

Senator CLINTON. What does that mean? If you have an on-scene coordinator, what level of responsibility does that suggest?

Administrator WHITMAN. If we were to determine that the cleanup was not thorough enough, that the capabilities did not exist with the on-scene coordinator or whomever was doing the actual decontamination, then we could go to court to supercede them, to come in and oversee the actual decontamination. There is an entire protocol of response, as you know, that the first people on the scene usually are the local responders and they are the ones who are the responsible party for seeing it through to the end, working very closely with the Federal Government. They bring in, depending on who is needed, the National Response Team and that is made up of all the different agencies. So it depends on what the threat is, what the problem is, who is involved in it.

Senator CLINTON. But you are not involved in Morgan Station which is the very large postal facility in Manhattan at all?

[Audience response.]

Administrator WHITMAN. If you could hear that answer?

Senator CLINTON. Yes, I did. Let me ask too about the protocol for dealing with the waste that comes out of these buildings after they are decontaminated. There was an article about Tom Brokaw's desk being sent to some waste disposal facility. Is there an existing protocol yet about what we do with the hazmat suits, with the vacuum cleaning equipment, with desks or other pieces of furniture or carpeting that has been infected? Is the waste going to a hazardous waste facility or where are we in the process of figuring out what we do with this?

Administrator WHITMAN. It is all treated as hazardous waste and taken to appropriate hazardous waste disposal facilities.

Senator CLINTON. Even if you are not involved?

Administrator WHITMAN. We have an on-scene coordinator who will make sure that nothing is overlooked in terms of the public safety. We don't make the final decisions. That is why Senator Corzine was looking for kind of a set pattern of how it happens.

There is really a set pattern, there is just not a set response. What we will do is have an on-scene coordinator, once there has been a determination that there is a biological agent that poses a threat to human health, we will have an on-scene coordinator onsite overseeing to make sure that those handling the actual decontamination are handling that appropriately and that the waste is handled appropriately as well. We can't force them to do things. We don't have the legal authority to force them to take actions. We can suggest, we can recommend, but the only way we could force is if we were to go to court to supercede them as the primary responder.

Senator CLINTON. So a city could make its own decisions, a county. Could a private facility basically decide they were going to handle it on their own and if there were another incident in New York, you could have a private company say they didn't want the New York City first responders, they didn't want the EPA, they were going to do this all by themselves? That is all permissible under the law as it currently stands?

Administrator WHITMAN. They could make that determination but if it was determined in fact there was an agent there, a biological or chemical agent that posed a risk to human health, we would have an on-scene coordinator and if we felt they were doing something to jeopardize people, then we could go to court to supercede them but no, you're correct in saying they have the primary responsibility for decisionmaking.

Senator CLINTON. It is clear we have a lot of work to do thinking this through, Mr. Chairman. I thank you for having this hearing.

Senator JEFFORDS. Thank you. You give me great confidence when you testify that you are doing the things that ought to be done. I appreciate very much working with you.

Administrator WHITMAN. Thank you very much, Mr. Chairman.

Senator JEFFORDS. In our second panel, we are fortunate to have two areas of expertise. Our first two witnesses will discuss the individual and community health concerns related to bioterrorism. Our second two witnesses will discuss various remediation technologies. I am hopeful we can glean lessons from our current ordeal by drawing on the depth of experience these four individuals offer.

First, we are pleased to have with us today, Dr. Patrick Meehan, Director, Division of Emergency and Environmental Health, Centers for Disease Control. He will discuss the health risks of original contaminated residuals after remediation, the health effects of decontamination, the remedy actions and the continued health monitoring requirements.

Dr. Meehan.

STATEMENT OF PATRICK MEEHAN, M.D., DIRECTOR, DIVISION OF EMERGENCY AND ENVIRONMENTAL HEALTH SERVICES, NATIONAL CENTER FOR ENVIRONMENTAL HEALTH, CENTERS FOR DISEASE CONTROL AND PREVENTION, DEPARTMENT OF HEALTH AND HUMAN SERVICES

Dr. MEEHAN. Good morning, Mr. Chairman, members of the committee.

As stated, I am Dr. Patrick Meehan, Director, Division of Emergency Environmental Health Services, Centers for Disease Control and Prevention. I want to thank you for the opportunity to discuss CDC's and the Agency for Toxic Disease, Toxic Substances, and Disease Registries which is ATSDR's role in support of the EPA in remediating anthrax-contaminated workplaces.

It is CDC's responsibility on behalf of the Department of Health and Human Services to provide national leadership in the public health and medical communities in a concerted effort to detect, diagnose, respond to, and prevent illnesses, including those that occur as a result of the deliberate release of biological, chemical, nuclear or radiologic agents. This task is an integral part of CDC's overall mission to monitor and protect the health of the U.S. population.

It is within this context that the CDC has begun to address preparing our nation's public health infrastructure to respond to potential, future and current acts of terrorism. Last year, CDC issued a strategy outlining steps for strengthening capacity to protect the Nation against threats of biological and chemical terrorism. This strategy identified five priority areas for planning efforts. A description of these areas has been provided in my written testimony which is submitted for the record.

Since the intentional release of anthrax spores, one of the areas in which CDC and ATSDR has focused is the identification and cleanup of contaminated facilities. We have refined methods for environmental sampling to assess whether anthrax contamination had occurred. In buildings that has meant sampling of air and/or surfaces. CDC and ATSDR have issued recommendations on how to conduct environmental sampling and how laboratories should analyze those samples. We also recommended environmental sampling strategies to characterize the extent of exposure and to guide cleanup.

We issued recommendations to protect first responders, investigators and cleanup personnel. As buildings were identified as contaminated, we provided technical input to EPA and others tasked with cleanup to determine where remediation was necessary. These recommendations have been widely disseminated to Federal, State and local health and environmental agencies and are available at CDC's bioterrorism website.

Disease experts at CDC are developing strategies to prevent the spread of disease during and after bioterrorist attacks. Although there is some data on chemical disinfectants in the scientific literature, there are no historical data that indicate the best way to eliminate spores from an office building or to disinfect a sorting machine as Senator Jeffords said in his opening statement.

The ability of a disinfectant to kill an anthrax spore is dependent upon time of contact and concentration and is mitigated by the amount and composition of material through which the disinfectant must penetrate to get to the spore. For many of the cleanup methods being used to kill anthrax spores, we will not know their effectiveness until we go through the process. EPA understands this and has sought help from a variety of sources including CDC and ATSDR to ensure that the appropriate indicators are used and that post-sampling strategies are adequate.

With regard to the effectiveness of cleaning, even our most exhaustive sampling strategies will not identify every spore. It is unlikely that any cleaning strategy will kill every spore. However, the EPA should be able to clean and retest to the point where we are all comfortable that spores have been killed or removed from surfaces where human contact is likely to occur. A range of sampling methods and strategies should be used to ensure the safety of building occupants.

In heavily contaminated areas such as Senator Daschle's suite and the Brentwood postal facility, fumigation is being proposed or has been used as the method of cleanup—the use of fumigants as a potential hazard for cleanup workers, those in areas adjacent to the buildings, and those that must reoccupy the buildings. A fumigant that is effective in killing spores is of necessity a highly toxic agent. The protection of workers during the fumigation process is a matter of good, industrial hygiene. EPA, CDC and ASTDR are working together to ensure remediation workers are protected during the fumigation process.

EPA works with local public health agencies to ensure that people in the area but outside of the building being fumigated are notified and kept at a safe distance. With regard to the safety of those who will reoccupy the building, it is important to determine both that the area is clear of fumigant and that there is no residual health risk. Again, CDC, ATSDR and the Occupational Safety and Health Administration have developed exposure limits for fumigants and detection methods are available to determine when any residual fumigant is well below established limits. After buildings are cleaned and post-cleaning environmental sampling has been conducted, CDC and ATSDR are committed to providing technical input to EPA and other experts to determine whether the building is ready for reoccupancy.

As highlighted recently, increased vigilance and preparedness for unexplained illnesses and injuries are an essential part of the public health effort to protect the American people against bioterrorism. Prior to the September 11 attack on the United States, CDC was making substantial progress toward defining, developing and implementing a nationwide public health response network to increase the capacity of public officials, to prepare for and respond to deliberate attacks on the health of our citizens.

The events of September 11 were a defining moment for all of us and since then we have dramatically increased all levels of preparedness and are implementing plans to increase them even further.

In conclusion, the best public health strategy to protect the health of civilians against biological and chemical terrorism is the development, organization and strengthening of public health surveillance and prevention systems and tools. Priorities include improved public health laboratory capacity, increased surveillance and outbreak investigation capacity, and health communication, education and training at the Federal, State and local levels.

Not only will this approach ensure that we are prepared for deliberate terrorist threats, but it will also improve our national capacity to promptly detect and control diseases not related to terrorism. A strong and flexible public health infrastructure is the best defense against any disease outbreak.

Thank you very much for the opportunity to speak today and I will be happy to answer questions.

Senator JEFFORDS. Thank you, Doctor.

Next we welcome Dr. Ivan Walks, director, Department of Health, Washington, DC. He will discuss the community health concerns related to bioterrorism and how the Federal Government can work better to inform and coordinate with local officials. Please proceed.

STATEMENT OF IVAN WALKS, M.D., DIRECTOR, DISTRICT OF COLUMBIA DEPARTMENT OF HEALTH

Dr. WALKS. Good morning, Chairman Jeffords and distinguished members of the Committee on the Environment and Public Works. I am Dr. Ivan C.A. Walks, chief health officer of the District of Columbia and director of the Department of Health. With me today is Theodore Gordon, chief operating officer, Department of Health, and key staff members involved with the remediation of biologically and chemically contaminated buildings.

We appreciate the opportunity to testify and commend you for convening this hearing because the discussion here this morning further complements our efforts to illuminate the issues regarding environmental exposures to contamination in the District of Columbia. This hearing also enhances our effort to continuously inform the community and involve them in decisions or procedures designed to address their concerns.

As I mentioned in the hearing on Spring Valley before the House of Representatives in July 2001, we cannot overemphasize the importance of an ongoing interaction between the government and the community. There can be no substitute for an informed community. That theme has been and will continue to be a guiding light for our efforts in every community in the District and in any other effort to prevent disease, dysfunction and premature death.

Allow me now to turn to the purpose of this hearing, i.e., the process that the District Government is guided in remediating biologically and chemically contaminated buildings and its progress and successes to date. My testimony will also cover the challenges that confront the District and the rest of the country, the new technologies available, and our next steps.

The Department of Health is relatively unique in that the Environmental Health Administration is part of our Department of Health and as such, we are charged with the mission of protecting human health via the prevention and control of environmentally related diseases, the prevention of environmental degradation and the promotion and preservation of the ecosystem and physical environment in the District.

When carrying out this charge, it is imperative that we follow a process that is structured but at the same time flexible enough to allow for stakeholder input. In this regard, and particularly with regard to time critical remediation, our process is similar to that described in the EPA's Superfund Community Involvement Handbook.

The District's process of remediation has as a first step identifying and defining the problem. Regarding biological contamination, this step involves both identification of contaminated regions of a building and all the possible pathways by which contamination can move beyond the contaminated zone to other locations within the building.

One of the things that I just discovered as we prepared for this hearing, because we are the Health Department, reportable diseases have to be reported to us. If a person is confirmed with anthrax, by law, they have to tell the Health Department. If a building is found to be contaminated with anthrax, that does not have to be reported to the Department of Health. That is one of those things that kind of makes you go "hmmm," and we really need to address that as an issue. I don't know if other jurisdictions have the same concern but certainly, it is one that we would like to raise at this hearing. Certainly, a contaminated building could lead to obvious concerns.

After that first step, we then begin to explore various remediation options. Each option is evaluated with regard to its technical effectiveness, practical feasibility and the unintended health and ecological risks to remediation workers and the adjacent community. For example, with respect to the Hart Building contamination, when we looked at the process being proposed, we looked at possible community contamination from leakage. Where there is an environmental standard for workers for a 15-minute exposure of 300 ppb, we advised a leakage standard of 15 as opposed to 300 but actually 15 ppb exposure and set up air handling and air monitoring facilities around the Hart Building so that if there was leakage, the Department of Health would be able to be involved and would step in.

I think our approach tends to be a bit more aggressive and maybe we can do that because we are a local health department but we certainly are very concerned and do our own monitoring when we are concerned about potential community risk.

In conducting an environmental risk assessment, several things are considered. First, we must be confident that we achieve a successful outcome. With regard to each option, we also have to consider costs, exposure to the government, community hardship emotionally as well as physically, and length of time for the cleanup. We continue to monitor, reevaluate throughout the planning and implementation stages. From all of those steps a prime option is then identified and we also focus on a secondary or fallback option so that we don't have to start from the beginning in case the prime option is not selected. Once we are almost certain we have considered all pertinent factors, we then prepare the plan of action, take it to the stakeholders for input and buy-in. We learned a long time ago that there is no such thing as a successful plan if the community doesn't help make the plan.

A big reason for our success in the Spring Valley community had to do with the inclusion of that community in our remediation strategy. We have had several meetings in the community, briefing its residents on our findings and process for remediation. In addition, Mayor Anthony Williams assembled an independent group, the Spring Valley Scientific Advisory Panel, which includes seven specialists in the fields of epidemiology, toxicology and environmental health as well as two representatives from the Spring Valley community.

The Department of Health has had significant experience in remediating biologically and chemically contaminated buildings in the District. Within the most recent 18 months we have experienced Legionella contamination in a correctional facility, a public school and a health-care facility. We have had significant fungal contamination of private homes and a public high school following a flood this past summer. In one community, private homes and the District Building were affected by a petroleum spill. Our successes are largely attributable to how well we communicate with the effected parties. Of course, we have a highly skilled and professional group of scientists and engineers who perform the technical risk assessment and remediation steps discussed above.

However, I must continue to stress the importance of communication as a key ingredient in any successful remediation plan.

There are several challenges confronting the District and the country. A particular challenge is that all health departments across the Nation regarding biological decontamination of buildings, is that these remediations must necessarily take place in a context of emerging science. We are all traveling steep learning curves with respect to the technical and medical facts. When we use toxic chemicals to kill biological agents, the scope of that learning curve must include stakeholders, both within and adjacent to the affected locations. In this regard, we wish to recommend one fundamental public health principle—until we learn whether a clinically significant minimum microbacterial contamination level exists—in other words, what does a little bit of anthrax mean—we should only declare a building to have been decontaminated when all test samples achieve nondetection levels. I think that is consistent with Administrator Whitman's earlier testimony.

With regard to community exposure to toxic chemicals, we must continue to maintain substantial margins of safety with regard to exposures to people in adjacent communities.

With respect to next steps, as we proceed to climb these steep learning curves, we need to share information real time with other State and local agencies. Such information must include biological sampling protocols, dosing, measuring, critical bioload levels and most of all, effectiveness data. We should expect the emergence of new chemical decontamination methods, rapid measuring technologies and biological detection methods. Knowledge of their efficacies and protocols should be widely shared within the public health community.

Thank you for this opportunity to come before you and discuss this issue and we would be happy to answer questions.

Senator JEFFORDS. Thank you, Dr. Walks.

Our next witness is Mr. Mike Grosser, technical director, Nuclear Biological and Chemical Defense Systems, Marine Corps Systems Command, with a long history of work in the decontamination field. We can learn a great deal from the Marines.

Thank you for sharing with us today the technologies that you have been investigating.

STATEMENT OF MIKE GROSSER, TECHNICAL DIRECTOR, NUCLEAR BIOLOGIC AND CHEMICAL DEFENSE SYSTEMS, MARINE CORPS SYSTEMS COMMAND

Mr. GROSSER. Good morning.

I am pleased to appear before you today to discuss several decontamination technologies that the Marine Corps and the Joint ChemBio Defense community have been pursuing. I am responsible to the program manager for the oversight of these programs. Although I am not a scientist, I have knowledge of the origin, the progress and the current status of them.

The Marine Corps has pursued these particular technologies as possible solutions to a requirement for an environmentally benign, patient-friendly, and effective personnel and equipment decontamination method. We did not set out to identify a specific decontaminant for anthrax-contaminated buildings.

The technologies I will talk about are by and large still in research and development. They have been considered as candidates for the joint service family of decontaminating systems and as tools for use by the Marine Corps' ChemBio Incident Response Force. While it is possible that one or two of them may be made available quickly, each has some facet that still requires research, testing and evaluation.

I will discuss electrochemically activated solutions, electrostatic decontamination, reactive nanoparticle technologies and the Sandia National Laboratory's foam decontaminant.

Electrochemically activated solutions are ECASOL which was developed in Russia in 1978. The Marine Corps has worked with electrochemical technologies at Las Vegas, NV and the Memorial Institute of Columbus, OH to further this technology. ECASOL is a colorless, odorless, aqueous solution, a mixture of water and salt that passes through a flow through electric nodule. The end product then is a decontaminant.

ECASOL is a highly effective biocidal agent. It is essentially a hypochlorous acid which is a close chemical relative of bleach. The ECASOL device developed for testing by the Marine Corps could be used to conduct a test for room and building decontamination, to conduct proof of principle type work and to see whether or not we have to apply liquid solution or if we can aerosolize that product. It is a highly effective anthrax killer in the laboratory in developmental tests but requires further testing with regard to application and to operational effectiveness.

Electrostatic decontamination is currently under development at the University of Missouri. This research and development program was started in 1998 and essentially we have an electrostatically charged mist containing a propriety photosensitizer that is sprayed onto a contaminated surface and then illuminated with a pulse UV light source. The photosensitizer mist will not cause injury to humans or damage to the environment.

I would like to note that it has not been developed or evaluated as a room or ductwork decontaminant but rather as a surface decontaminant. We believe that ESD could be misted into enclosed spaces, possibly ductwork to effectively neutralize biological agents. This developmental effort would require some minor modification of commercial, off the shelf technologies, some applicators and testing to ensure that proper procedures are in place to maximize agent neutralization.

Reactive nanoparticle technologies involve a nanoparticle regime that includes materials with particle sizes ranging between 1–100 nanometers. Nanoparticles of metal oxide exhibit extraordinary abilities to react with and thereby destroy highly toxic substances and chemical warfare agents.

Kansas State University and a commercial adjunct firm, Nanoscale Materials Inc., have been active since 1995 in developing metal oxide nanoparticles. Since August of this year, the Marine Corps Systems Command has aggressively pursued this technology for a wide range of decontamination applications. This project is focused on developing a novel dry powder decontamination technology capable of neutralizing chemical and biological warfare agents in the effort to get away from the aqueous decontaminant.

It has shown some promise lately as a biological killer. This technology could be available for use as early as calendar year 2003 given the appropriate resources.

Sandia National Laboratory has developed the fourth technology. This decontaminant designated DF–100 is nontoxic, noncorrosive, an aqueous foam with enhanced physical stability for the rapid mitigation and decontamination of chemical and biological warfare agents.

I believe I heard earlier today that a similar foam has been used in remediation efforts in congressional office buildings, although I don't know how effective those efforts have been yet. The foam formulation is based on a surfactant system, a solubilized decontaminant. The formulation includes water soluble powders to enhance the physical stability of the foam.

Preliminary test results demonstrate very effective decontamination of chemical and biological agents, something heretofore not seen by us. The decontaminants I mentioned earlier are largely biological killers.

The decontamination technology may offer the following benefits. We believe it could be a single decontaminant solution for both chemical and biological threats, it may be rapidly deployed and has a minimal operational logistic impact.

Decontamination demonstrations at Dugway Proving Ground, UT and Ft. Leonard Wood have shown that DF–100 may be applied with currently field decontamination systems such as firefighting equipment or even pressure washers.

In conclusion, I want to thank the committee for inviting me to present this information. The Marine Corps and the Joint Chem-Bio Defense Program continues to conduct research and development and acquisition of all these technologies.

I would be happy to answer any questions at this time.

Senator JEFFORDS. Thank you again for excellent testimony in an area in which we desperately need information.

Our final witness, we will hear from a company in an industry that is sure to grow in the near future. Mr. Les Vinney is president and CEO of the Steris Corporation, a provider of technologies for infection and contamination prevention. Welcome. We are looking forward to the answers.

STATEMENT OF LES VINNEY, PRESIDENT AND CEO, STERIS CORPORATION ACCOMPANIED BY: PETER BURKE, VICE PRESIDENT AND CHIEF TECHNOLOGY OFFICER; GERRY REIS, VICE PRESIDENT FOR CORPORATE ADMINISTRATION; KARLA PERRI, SENIOR ENVIRONMENTAL CONSULTANT, VERSAR, INC.

Mr. VINNEY. Thank you. Good morning, Mr. Chairman and members of the committee.

Thank you for your invitation and welcome the opportunity to address you on this critically important issue. I would request that the formal written statement we provided be submitted for the record.

I am accompanied this morning by Dr. Peter Burke, vice president and chief technology officer and by Mr. Gerry Reis, vice president for Corporate Administration. Also joining me is Ms. Karla Perri, senior environmental consultant of Versar, Inc.

Steris Corporation technologies are used every day in environments where the highest levels of sterility are required. Health care professionals in virtually every U.S. hospital and researchers, scientists and the pharmaceutical industry, including all of the Fortune 50 pharmaceutical companies use Steris products to sterilize and decontaminate items from surgical instruments to their equipment and facilities. These technologies help ensure positive outcomes of such critical activities as the production of antibiotics, the development of vaccines and the safety of sensitive medical devices and implants for humans.

The primary business focus of Steris is to develop and produce formulations that prevent infection and contamination and the delivery systems to enable their most efficient use. When properly utilized these technologies can provide safe and effective remediation of contaminated materials in whatever form they may take including entire rooms and their contents. These technologies can also be put in place to prevent recontamination and assure ongoing safety, as is their purpose in the industries we currently serve.

In light of the recent events in our country, we welcome the opportunity to offer our expertise to help restore biologically contaminated facilities for normal use. We believe that our technologies can help to optimize and improve the safety of remediation efforts both in their application and potential residual effects. Toward that end, we have joined with Versar, Inc., a leader in providing counterterrorism, environmental, architectural, engineering and related services. Together, Steris and Versar offer a broad array of contamination risk assessment and remediation services.

We firmly believe that methods now in use in health care and scientific settings can effectively decontaminate facilities infected with anthrax. The reason you have not previously seen us before your committee is that the large majority of Steris products are traditionally used in hospitals and by pharmaceutical companies. As such, we normally have had or technologies and processes accepted for use under the purview of the Food and Drug Administration. While many of our formulations have been registered for specific uses with the EPA, our decontamination processes have not previously been registered for such applications as mail and building decontamination of the kind our Nation is now addressing.

While our past experience gives us very high confidence in the effectiveness of our technologies, we strongly endorse the regulatory requirements to test and validate a product technology prior to allowing its use for specific treatment applications. In that regard, we have been seeking the opportunity to demonstrate the efficacy of our product technologies to meet various remediation needs.

As no bridge exists across regulatory jurisdictions to enable the more rapid application of these existing capabilities to meet emergency decontamination requirements, we have had to develop new working relationships for this purpose over the last several weeks. We are now working closely with the EPA to secure the necessary approvals to permit the use of our technologies for these applications. We are also in advanced discussions with the Department of Justice on a potential demonstration project which would serve to validate the effectiveness of our technologies in decontaminating anthrax-infected facilities.

The health care and pharmaceutical industries have dealt with microbial control challenges for many years. As a result, highly sophisticated prevention and treatment methodologies have been de-

81-720 D-3

veloped within these industries. While older technologies such as formaldehyde and chlorine dioxide have been used in these industries, more technologies such as vapor hydrogen peroxide and a combination of hydrogen peroxide and parcetic acid sporicidal compounds have been developed. These emerging technologies have displaced the earlier technologies because they offer certain advantages—reduced toxicity, limited corrosiveness, minimal residual effects and easier application.

A facility contaminated by highly aerosolized anthrax spores offers a unique and severe challenge. While these conditions present a different environment than our more standard applications, to accomplish proper remediation a carefully planned approach would be used similar to those followed in establishing the preventive process for health care and scientific requirements.

In an appendix attached to my written testimony, we have presented a detailed plan for systematic biological remediation of a given facility or area.

Mr. Chairman, in our profession view there is no single silver bullet for treating biological contamination. This remediation requires the selective use of multiple technologies. This approach should result in the least damage to items within contaminated facilities, assure that each surface and material is treated with the agent best suited to its individual needs and therefore, provide the highest level of decontamination.

In closing, we believe a coordinated effort is needed among appropriate government, academic, military and private industry officials. This coordinated approach will permit the identification, validation and utilization of the safest and most effective technologies currently available. Careful development of the proper protocols for this remediation process is critical to a successful outcome.

What we must achieve is the restoration and maintenance of safe working environments for all Americans. Steris stands ready to help.

Thank you for the opportunity to appear before you today and I would be happy to answer any questions you may have.

Senator JEFFORDS. Thank you, all of you. This has been extremely helpful. I have much more confidence as we finish with your testimony that we are going to make progress and be able to get things under control. I do have a question or two for each of you.

Dr. Meehan, how does Hart's current decontamination plan take into account the greater level of risk to those individuals who are allergic to, pregnant or have compromised immune systems and cannot take the antibiotics used to treat anthrax and/or be more sensitive to the decontamination remedy?

Dr. MEEHAN. Regarding the antibiotic treatment, antibiotic prophylaxis of the people who were exposed, CDC has published very clear recommendations about what antibiotic regimens should be used including appropriate treatments for children, pregnant women and in the unusual situation in the Capitol Hill physician who is overseeing the treatment of the staff in the Senate building, has worked very closely with us. We have been in close contact with him as well and we have been present in the operations center for the Capitol Hill project right from the beginning, so there has been extensive give and take and involvement, very clear guidelines on what antibiotic regimens are appropriate, how to deal with children and pregnant women.

In the unusual case of antibiotic sensitivity, those need to be dealt with on a case-by-case basis and we are available to consult with the Capitol Hill medical staff on a case-by-case basis for those.

Regarding the fumigant that is used, as I stated in my testimony, there are very clear guidelines for exposure limits for workers and the general population published by both NIOSH and OSHA. We will continue to work with EPA to assure that no one is exposed to any level that would be a health risk to anyone as they reenter the building.

Senator JEFFORDS. What steps has CDC taken to educate doctors about anthrax and what type of system is in place to educate health providers as quickly as possible in the case of another biological or chemical attack?

Dr. MEEHAN. Prior September 11, we had a bioterrorism preparedness program that had been in development for about 2 years. We have projects with every State health department, plus the District of Columbia and others, New York City in particular. During that time we had extensive efforts to work through the health departments to educate the medical community about bioterrorism agents and how to detect and treat them. Since September 11, we have really revved that up. We have had a number of satellite video conferences targeted to clinicians, collaborated with the AMA and others on that. We have, through our Health Alert Network, a computerized, Internet-based system that goes down to the provider level, almost on a daily basis provided updates to health departments and clinicians and we have had an operations center at CDC open 24 hours a day, 7 days a week and have gotten literally thousands of phone calls from providers out in the community about patient questions. "Could this be an anthrax case?" "What sort of medical test should I do?"—that sort of thing. I think the medical community is aware of our availability.

Senator JEFFORDS. In your testimony, you mentioned the need to ensure that post-sampling strategies are adequate. What type of monitoring will you be doing after remediated buildings have been reoccupied?

Dr. MEEHAN. Prior to reoccupation, we will work with EPA to do extensive resampling to assure there are no detectable spores that would be of any risk to anybody which means essentially no detectable spores.

After folks reenter the building, we will continue to do what we have been doing in every community that has had an anthrax attack. That is to work with the local medical community and the health department to aggressively monitor the population for possible anthrax-related disease. We work with the health departments and thereby work with the clinicians and the community so that if there is a patient that might potentially have anthrax, they are reported immediately to the health department and that particular case is investigated so we can catch any cases early. We are optimistic that wouldn't occur, however, but there is a system in place for doing surveillance and making sure we catch any cases as early as possible.

Senator JEFFORDS. Dr.Walks, I agree with you that the community involvement is a critical component of the emerging science known as building remediation. I think we have a unique challenge of addressing an emergency situation while ensuring involvement of all stakeholders. I am interested in your thoughts as to how we can improve community stakeholder involvement in building remediation efforts?

Dr. WALKS. I think we have some good examples here locally. It is unfortunate but maybe useful that this has hit the Nation's Capitol because a lot of folks have had a chance to see what has happened here. I think during a time of crisis having routine, regular, dependable communication with the public helped. Dr. Meehan certainly played a terrific role in working closely with us. Our Mayor showed tremendous leadership in being out front and routinely and regularly giving folks real time information.

With respect to the Hart Building cleanup, though we have multiple jurisdictions in the District, if it is the White House, it is Dr. Tubb; if it is the Capitol, it is Dr. Reishold; if it is the District, it is Dr. Walks; but we all seem to communicate pretty well despite that and I think maybe because of that communication, we were able over a week ago to have a community meeting to involve the community prior to the Hart Building remediation and let folks know what was going to happen and to take on another challenge which is communication across diverse communities.

When we have something come to an urban area like the District, like New York, like Chicago, Los Angeles or Atlanta, you are talking about multiple languages, over 120 languages in Los Angeles, nearly 100 languages spoken here in the District. If we don't find a way to communicate across cultures and across linguistic barriers, then we really don't involve the community. I think examples of community meetings before plans are finalized, examples of talking across jurisdictions, talking across cultures is really how we do that.

Senator JEFFORDS. What lessons have you as Director of the District of Columbia Department of Health gleaned from this experience with bioterrorism?

Dr. WALKS. I think the overriding lesson learned is the tremendous spirit of cooperation and focus that we have seen. I have said this previously but the CDC was in the District ready to work on the ground within 3 hours of the first confirmation that the Daschle letter contained anthrax. I think that is remarkable. I think what we have seen is a community that has shown tremendous resiliency. I think we have seen leadership that is critical.

Maybe the most important lesson learned is early, clear leadership. I think we have seen Mayor Williams locally, Mayor Guiliani in New York, stand up in front of the cameras and tell people what they don't know so folks will believe you when you tell them what you do know. I think that is probably the most important lesson learned, that people can respect honest ignorance. If you don't know and you tell people that, you build a relationship based on honesty. If we do anything going forward across the country, we need to trust people with information, trust that people are not going to panic. We are a resilient culture across the board and if we know what is coming, we know what to expect, then we can respond appropriately and work together. We saw that and the jurisdictions meant nothing.

The first press conference we had locally about the anthrax was a regional press conference involving the Secretary of Health, Georges Benjamin from Maryland, Ann Peterson, who was the Commissioner of Health for the State of Virginia and I think those sorts of cross jurisdictional efforts are going to be critical.

To sum Chapter 1, clear leadership, cross jurisdictional cooperation, honest information, timely information, I think those are probably the biggest lessons learned.

Senator JEFFORDS. Mr. Grosser, I am intrigued by your testimony. I don't think many of us thought that you obviously worried about these kinds of things for years in the military, warfare and all. Which technology you discussed in your testimony is least harmful to human health and the environment?

Mr. GROSSER. Remember, sir, all these technologies are still in research and development so I need to couch ahead of time that things may emerge here in the near or mid term that contradict what I am about to say. I believe we know that ecosol, the first technology I discussed, is harmless to human beings and that the effluent, the runoff, after it is decontaminated, is environmentally benign, so I would say certainly ecosol.

I do not know about electrostatic decontamination. I know that it is harmless and the photosensitizer is harmless, the UV light is harmless but again we haven't tested these operationally, only developmentally.

Electrostatic particles are still a ways in development and Sandia foam, I think the EPA is better at answering that than I. They have more experience with that here recently.

Senator JEFFORDS. If ECASOL was developed for personal decontamination and skin contact, what makes you think it will work on a building and has this ever been used or something like that on a building?

Mr. GROSSER. I am not aware that it has ever been used or anything like that on a building. What makes us think it would work is it is an extremely effective anthrax killer. We are pretty innovative in coming up with implications, operational techniques to make decontaminants work, both for personnel and equipment. I don't know what that method would be but given the opportunity, I think we could develop some operational tests, some protocols to determine just how effective it might be.

Senator JEFFORDS. Mr. Vinney, do you agree with the EPA's determination that chlorine dioxide was the best treatment to proceed with in the case of the Hart Senate Office Building?

Mr. VINNEY. Certainly the Federal agencies responded to the crisis using what they know and with technologies that have worked in the past. I think the situation we are dealing with is unprecedented, of course, and being biological contaminants, the treatment of biological contaminants is really not new but is handled in a very different setting and that is in the medical and pharmaceutical environments.

The technologies the EPA is using today were used by the pharmaceutical and medical environments previously but there have been advances and they have moved on to products such as vapor hydrogen peroxide and others. I think there is certainly appropriate situations in which chlorine dioxide and the other types of materials that are being used would be used but you would have to examine the specific situation and the environment in which it was going to be used and we have not had that opportunity obviously.

Senator JEFFORDS. I want to thank all of you for your very, very helpful testimony. We reserve the right to continue to grill you through the mail, so don't get too relaxed but this has been very helpful to the committee and I thank you for the effort and time you have put into the presentations and for answering the questions. Thank you.

[Whereupon, at 11:18 a.m., the committee was adjourned, to reconvene at the call of the chair.]

[Additional statements submitted for the record follow:]

STATEMENT OF CHRISTINE TODD WHITMAN, ADMINISTRATOR, U.S. ENVIRONMENTAL PROTECTION

Chairman Jeffords and members of the subcommittee, thank you for the opportunity to describe the Environmental Protection Agency's (EPA) role in combating bioterrorism: specifically, the role in the decontamination of anthrax in buildings as part of the Agency's overall mission to protect human health and the environment. I am pleased to say that EPA's efforts to meet its counterterrorism obligations are consistent with the President's statement that combating terrorism and protecting the nation's critical infrastructures are a high priority for his administration.

There are several Presidential Decision Directives (PDDs) that specify a role for EPA in counterterrorism activities. PDD 39 assigned EPA the task of assisting the FBI during crisis management in threat assessments and determining the type of hazards associated with releases or potential releases of materials in a terrorist incident. EPA, as the lead agency for Hazardous Materials Response under Emergency Support Function (ESF) 10 of the Federal Response Plan, is also assigned to assist the Federal Emergency Management Agency, during consequence management with environmental monitoring, decontamination, and long-term site cleanup. PDD 62 reinforces our mission to enhance the nation's capabilities to respond to terrorist events. PDD 63 which addresses the protection of America's critical infrastructure, named EPA the lead agency for the Water Supply Sector.

Working with our Federal partners, private sector experts, and drawing upon our considerable in-house expertise, EPA has been developing new methods and protocols, and standard operating procedures to deal with this new threat to the health and safety of the American people. And we have been doing so on a real-time basis. The speed of our response, however, has not been at the expense of sound science. Indeed, a team of science experts has been integral to our daily activities.

EPA'S ROLE IN BUILDINGS CONTAMINATED WITH ANTHRAX

Our cleanup experts have been drawing on their years of expertise and experience, on the talents of scientists in industry and academia, and on the knowledge available from our Federal partners. Similar analysis informed the cleanups undertaken at the several postal facilities and media offices, although since they were of a much smaller scope, they were more readily addressed.

Our role at a site generally begins after the Centers for Disease Control and Prevention (CDC) has tested to determine the presence of a threat and the risk that threat poses to human health. Once a decision is made to decontaminate a building, EPA and CDC will work together to advise the Incident Commander about the extent to which a building must be cleaned to make it safe.

EPA staff has provided expert technical advice to facility managers throughout the country on issues such as sampling plans, worker safety and actual site cleanup methods.

This role is a natural fit for EPA's on-scene coordinators, managers who are experienced in assessing contamination in structures, soil, water and air-handling systems. On-scene coordinators have considerable experience at sorting out hazards, quantifying risks, planning and implementing emergency cleanups, and coordinating among other agencies, State and local government, and the private sector.

EPA employees are working at the direction of the incident commanders from other Federal agencies, and report to the U.S. Postal Service at their facilities and the Sergeant at Arms in the Capitol.

In addition to the activity generated by testing and cleaning, these sites are also being treated as crime scenes. That is why our Criminal Investigation Division has been working closely with the FBI and with local and State law enforcement agencies at the various contaminated sites. We are assisting the FBI in gathering evidence to identify the criminals responsible for terrorist attacks.

As we seek to apply the lessons we're learning from all our decontamination efforts one thing is becoming clear—there's no one-size-fits-all solution. Each event has to be thoroughly analyzed as a separate case before we can propose an effective solution.

For example, cleaning a facility that largely contains rugged, heavy equipment can be accomplished using such methods as foam or liquid chlorine dioxide—methods that the contents of the building can stand up to. On the other hand, a facility that contains lots of paper, office furniture, and electronic equipment needs to be cleaned using another method—such as fumigation—that won't damage the contents in the way a liquid would.

Other factors, such as the amount of contamination found, the ways and extent to which it can be dispersed throughout a building, the nature of the surrounding area, and the ways in which the building is used all require additional consideration before proceeding with decontamination.

The first step in remediating a building is just like the first step in any cleanup operation and that is to determine the potential for risk to human health. Anthrax is a known threat to human health, but the literature is scant on the number of spores that a person must be exposed to before developing inhalational disease.

The health team that has come together to help us establish the parameters for defining the extent of contamination and providing direct health advice to affected individuals has involved a wide array of experts. The Congress's own Office of the Attending Physician has played a central role in providing direct medical advice to the people who work in the affected buildings. EPA has worked with the CDC and the Agency for Toxic Substances and Disease Registry in the Department of Health and Human Services in the areas of sampling strategy, remediating processes and criteria for judging a remediation process to be effective. In particular, National Institute for Occupational Safety and Health (NIOSH) within CDC has been extremely helpful as has been the Department for Labor's Occupational Safety and Health Administration (OSHA). The Department of Defense, including the U.S. Air Force's CHPPM group has special expertise because of the potential that anthrax would be used as a biological weapon in a war setting. OSHA has been helpful in determining appropriate safety measures both for the people who work in the buildings and also for the extensive remediation crews that are at work here. The District of Columbia's Department of Health as well as their State counterparts, Maryland's Department of Health and Mental Hygiene, have been consulted regularly. And EPA's own in-house expertise including toxicologists from as far away as our Denver office and safety officers from our own nearby Ft. Meade laboratory have also played a vital role.

Together this group of experts has reached consensus on when cleanup activities are warranted, and they have also formed a team to review final cleanup data to make a determination that the buildings will be safe to reoccupy.

REMEDIATION STRATEGIES

While we have developed extraordinarily strong working relationships with numerous partners in developing the appropriate health and safety standards and in conducting our sampling work, it is in the area of actual remediation efforts that our collaborations have been the most broad-based.

The full array of Federal agencies with expertise in remediation strategies has been involved in helping develop the tools we need to deal with anthrax contamination. We have consulted with the White House's Office of Science Technology Policy. Indeed, the President's science advisor has been at the Incident Command Center, providing a key link to this Federal Government-wide response.

Additionally, we are gratified by the level of cooperation and coordination that has taken place between the Federal agencies with responsibilities for identifying and remediating anthrax contamination. In particular, we have worked very closely with staff from the Centers for Disease Control and Prevention in the areas of sampling strategy, remediating processes and criteria for judging a remediation process to be effective. NIOSH has been extremely helpful in providing EPA expertise in the area of worker protection, both for response operations and in establishing cleanup goals.

We also appreciate the input from the Department of Defense, particularly the Center for Health Promotion and Preventive Medicine and U.S. Army Medical Research Institute for Infectious Diseases. The Coast Guard and Marines have assisted with sampling and cleanup. Finally, the District of Columbia government has provided invaluable expertise and assistance in involving the community.

At EPA, our Office of Solid Waste and Emergency Response, the Office of Pesticides, our Environmental Response Team out of Edison, NJ, the Emergency Operations Center here in Washington, and the legion of responders from across the country led by our folks from Region III, have all played important roles in the cleanup effort.

A number of liquid and foam applications are effective at actually killing spores. Sandia Foam is a patented product, developed by the Sandia Labs, that we have been able to use on a number of surfaces. Similarly, chlorine dioxide in a liquid form, has been an extremely effective sporocide. We know these techniques work because we have used them in a number of areas. To address airborne particles, HEPA (high efficiency particulate air) filter vacuums are able to capture particles down to less than one-half micron in size. After the remediation effort is complete, we have resampled these areas and they have come back clean.

The tools in our toolbox are growing rapidly. Each method, though, will have to prove its effectiveness before we add it to our Standard Operating Procedures. And that proof will come from confirmation samples that are taken after remediation is complete and come back demonstrating no threat to human health.

EPA'S COUNTERTERRORISM INCIDENT RESPONSE ACTIVITIES

As EPA continues to strengthen its counterterrorism (CT) program by building on the existing national response system for hazardous materials (hazmat) prevention, preparedness, and response, the Agency is involved in a variety of activities with Federal, State, and local officials that include: responding to terrorism threats; predeploying for special events; planning, coordination, and outreach; and training and exercises. Most recently, EPA was asked to chair the Security and Safety of U.S. Facilities Group of the National Security Council's Policy Coordinating Committee for Counterterrorism and National Preparedness.

EPA established and maintains a National Incident Coordination Team (NICT) to assure full agency coordination of all emergency preparedness and response activities including counterterrorism. In the regions, the Agency's first responders are the On-Scene Coordinators (or OSCs). The OSCs have been actively involved with local, State, and Federal authorities in preparing for and responding to threats of terrorism. EPA's OSCs, located throughout the United States, have broad response authority and a proven record of success in responding rapidly to emergency situations.

REGISTRATION OF PRODUCTS

Another principal responsibility of EPA's in anthrax decontamination is to ensure that the chemicals used to treat anthrax spores are efficacious and safe. EPA is responsible for registering pesticides, including these antimicrobial products used to treat anthrax spores, prior to their marketing in the United States.

Before issuing a pesticide registration, the Agency reviews a significant body of data to determine whether use of that pesticide will result in unreasonable adverse effects to humans or the environment. These data can include information on short- and long-term toxic effects and examine the potential for exposure under expected application scenarios. For pesticides that have public health uses, such as those used on anthrax spores, EPA also critically evaluates their efficacy. Under emergency conditions, EPA may allow a new use of a previously registered pesticide or use of an unregistered pesticide where the Agency has sufficient data to make a safety finding. These decisions can often be made quickly, based on the data that EPA receives and reviews.

Responding to the anthrax contamination has presented some unique challenges to our pesticides program. For example, currently there are no registered pesticides approved for use against anthrax. Since the beginning of the anthrax-contamination events, EPA has been working hard to identify and evaluate existing pesticide products that are sporicidal, that is, those that kill spore-forming bacteria, even though such products may not have been tested on anthrax per se.

Since October, the Agency has approved two pesticides for treating anthrax spores under emergency exemption provisions of existing pesticide laws—the aqueous solution of chlorine dioxide and a foam used to treat anthrax-contaminated surfaces. We have identified several potential chemicals and new technologies which may be effective against anthrax. The Agency continues to work closely with other Federal agencies, emergency response teams, and independent experts to develop effective remediation tools. On the basis of site specific information, EPA recommends proper methods of decontamination including which antimicrobial or other substances will be used.

EPA has also established a hotline for venders who believe they have products that could effectively treat anthrax and has begun daily briefings to establish routine communication between onsite personnel and key centers within the Agency who oversee and/or support them. EPA laboratories are assisting in testing samples from potentially contaminated sites and the evaluation of antimicrobial products for effectiveness against anthrax has been made a top priority. In addition, EPA is using its experience in this situation to develop approaches to handling future biological and chemical exposures should they occur.

CONCLUSION

September 11 has changed the world in which we live. EPA continues to rely on sound science and effective treatment techniques to address the threat of anthrax contamination in some of our Nation's buildings. We are proud to be a part of a massive public-private effort to meet the challenges of this new world.

Thank you for the opportunity to appear before you today. I would be happy to answer any questions that you may have.

RESPONSES BY HON. CHRISTINE TODD WHITMAN TO ADDITIONAL QUESTIONS FROM SENATOR JEFFORDS

Question 1. Who is accountable for building remediation? Is the Hart Senate Office Building a unique situation? Under what authority does EPA act to remediate buildings?

Response. EPA has authority under the Comprehensive Environmental Response, Compensation and liability Act (CERCLA) and an Executive Order establishing the National Oil and Hazardous Substances Contingency Plan (NCP), to clean up contamination within a building if the Agency believes that there is the possibility that the contamination may leave the building and create a release to the environment. As part of its efforts, EPA may deploy Federal resources to do monitoring, sampling, risk assessment, safety and health analysis, clean up, disposal, and other response requirements.

The EPA may defer to the owner/operator of a facility (whether private, Federal or State/local) to carry out a response to an incident involving hazardous substances and oil, if the Agency determines that they have appropriate response capability. In these instances, the EPA provides oversight and technical support. EPA has the authority to take over such responses, if necessary, in order to protect public health and the environment.

In the situation of the Hart Office Building, the U.S. Capitol Police and Sergeant at Arms have the lead in carrying out the appropriate response, and EPA is providing technical support.

Because EPA has never before conducted a response to a biological agent such as anthrax on this scale, the Hart Senate Office Building is a unique situation. The experience gained from the Hart cleanup has provided valuable guidance to the inter-agency National Response Team for any future anthrax response actions.

Question 2. At the hearing, you discussed an expedited procedure for approving remediation technologies. I am interested in hearing more about that. Which remediation technologies are currently on EPA's list, and what process will you employ to determine further technologies?

Response. The decontamination of anthrax is a rapidly evolving field, with new technologies continually being advanced and tested. EPA continues to receive information from vendors of potentially effective decontamination technologies and we are coordinating with other agencies to evaluate the effectiveness of these technologies in decontaminating anthrax. The Agency reviews such claims very carefully and places priority on those products that appear to be most promising for use in decontamination plans. The Agency also gives weight to requests by EPA On Scene Coordinators (such as those on Capitol Hill), by other Federal agencies, and by the U.S. Postal Service in determining which products are needed immediately for treatment of contaminated facilities.

Under section 18 of the Federal Insecticide, Fungicide and Rodenticide Act, EPA may temporarily exempt products from the registration requirements for uses which are deemed necessary under emergency situations. These exemptions are granted after considering available data and include requirements which ensure protection of health and the environment. EPA has provided these exemptions for several antimicrobial pesticides for use as part of decontamination plans under this expedited process, including chlorine dioxide, Envirofoam, and ethylene oxide.

At this time, EPA has been notified by and, in most cases, received information from vendors on the types of products listed below. EPA is proceeding to evaluate these products on an expedited basis. No product should be used until or unless EPA has approved or exempted it, and any product that is used should be applied in a comprehensive program that involves sampling, cleaning, treating and resampling followed by retreatment as necessary to ensure effective decontamination.

- Liquid antimicrobials: Bleach (sodium hypochlorite) liquid chlorine dioxide, hydrogen peroxide, hydrogen peroxide and peracetic acid, hydrogen peroxide and quaternary ammoniums, hydrogen peroxide and ethanol, hydrogen peroxide and silver, iodine, nanoemulsion, parachlorometaxylenol, phenolics, quaternary ammoniums, silyl ammoniums, isothiazolones, silver, and other proprietary mixtures.
- Gaseous antimicrobials: Gaseous and fogged chlorine dioxide, ethylene oxide, vaporized hydrogen peroxide, and vaporized paraformaldehyde.
- HEPA vacuuming
- Irradiation
- Pesticide devices: ozone generators, electrostatic systems, chemical/steam systems, and ultraviolet light and ultrasound.
- Destructive Disposal: Incineration, autoclaving

Question 3a. I understand that EPA's been doing ambient monitoring for fine and toxic air pollutants around the World Trade Center site. But, some health officials have reported that some of the most obvious acute health problems (eyes, nose, and throat irritation) associated with the larger, alkaline and caustic particles have been overlooked. How can we be sure to have a system in place that will monitor for all possible air pollutants in preparation for any possible future disasters/attacks?

Response. We can never be completely prepared for such attacks. Each incident will have unique aspects with respect to its investigation and evaluation. A more refined and specific approach, however, may assist us if such an event were to occur again. Specifically a phased-in monitoring approach, as used at the World Trade Center, may be a useful initial step. Most events consist of (1) an immediate response (for example, the visible particles in the air); (2) a steady state response (for example, you can see the plume from the fire) and (3) return to pre-incident conditions (for example, the fire is waning; people are returning to their homes and businesses). For each stage, different decisions are made as to the appropriate course of action and the next steps, and at each phase there could be different monitoring.

Question 3b. Could simple mechanisms like dust masks have assisted the public with their health concerns?

Response. Dust masks would have been helpful for particles. For gases, however, a different breathing apparatus and the appropriate filters would be needed to be effective, and the efficiency of protection would depend upon mask type and design.

Question 3c. Do you think it would be helpful to have a NAS panel review the environmental health risks associated with the World Trade Center site that we can use as a lesson for future incident preparation?

Response. A NAS panel review of the environmental and health impact issues surrounding the World Trade Center monitoring efforts would be useful.

Question 4. The committee has heard from individuals near the World Trade Center site about their concerns regarding asbestos contamination. The Federal Government has largely stayed away from setting indoor air quality standards. Should we be doing more to ensure that there are such standards, for public places at a minimum, so that people can be certain of some level of protectiveness across the country?

Response. The issue of setting national indoor air quality standards is a complex issue particularly in light of the sheer number and variability of indoor spaces. Traditionally, indoor air quality has been left to local authorities. The indoor environment has been considered to be outside the scope of the Clean Air Act's standard setting authority. All this being said, however, EPA believes that Congress and the Administration need to revisit the issue of authority and responsibility for indoor environmental conditions in the wake of a terrorist attack. It may be that the current practice, vesting in local and state governments primary responsibility for indoor environmental conditions is not appropriate in the wake of an event like September 11th.

Question 5. As you may know, the GAO is currently undertaking a congressionally mandated study to assess information that Local Emergency Planning Committees (LEPCs) receive from EPA and elsewhere in order to respond to chemical accidents and toxic releases. GAO's initial inquiries have turned up somewhat disturbing news. Many of the LEPCs don't function as effectively as they could due to lack of funding, and it is next to impossible to contact them in a coordinated fashion. Shouldn't EPA or FEMA have a reliable system of alerting these entities to national, regional or even local emergencies, especially if it might include a coordinated terrorist attack on chemical facilities? Is the lack of coordination simply a matter of resources? How much funding does EPA provide to the LEPCs? Does or should this money come from fees collected from those facilities that create the risk, i.e., chemical, petroleum, etc.?

Response. It is a challenge to distribute timely information during emergencies. To try and address this issue, EPA's current system relies on the assistance of the State Emergency Response Commissions (SERCs) that appoint the Local Emergency Planning Committees (LEPCs). When there is information that needs to be quickly disseminated, EPA communicates with the SERCs, who then provide information to each LEPC. For more routine communications, EPA regularly sends information update letters to State Commission chairs, and operates an Internet list-serve to which many LEPCs subscribe. EPA uses this list serve to provide LEPCs and other subscribers with chemical safety updates and news items. Following the events of September 11, 2001, EPA used this list serve to distribute a chemical site security advisory using this system. EPA has posted on its website a data base of LEPCs with contact information; members of the public can search for the appropriate LEPC contact information for their geographical location.

In the years immediately after Congress created LEPCs with the 1986 passage of the Emergency Planning and Community Right to Know Act (EPCRA), EPA allocated approximately $1 million each year of programmatic funds for LEPC grants. EPA made these grants using other statutory authorities, as EPCRA does not include grant authority for EPA to support LEPCs or SERCs. In the early 1990's, Congress amended the Hazardous Materials Transportation Act to provide the Department of Transportation the authority to charge transporters fees and to use those fees for training and planning grants for LEPCs and local responders. Some States have established a fee system under which fees from facilities are used to support LEPCs and the SERC, at different levels in different States.

RESPONSES BY HON. CHRISTINE TODD WHITMAN TO ADDITIONAL QUESTIONS FROM SENATOR CORZINE

Question 1a. I understand that bleach was used to clean up anthrax contamination at several New Jersey post offices, as opposed to the Sandia foam that is being used in the Hart building. Do you have any studies or data that indicates the relative effectiveness of these two decontaminants?

Response. EPA does not have data that specifically compare the effectiveness of bleach and Sandia foam in anthrax decontamination scenarios. EPA has reviewed available data on Sandia foam and concluded that it can be used effectively as part of an anthrax decontamination plan. EPA is currently conducting tests on the use of bleach in anthrax decontamination plans but has not yet reached definitive conclusions about the conditions under which it is effective. Under certain conditions (i.e., hard surfaces, specific pH, adequate contact time) bleach can be used in a decontamination plan but, as with all chemicals for use in anthrax decontamination, the cleanup and treatment process must be followed by a thorough post-treatment sampling of the contaminated area to ensure that effective decontamination has taken place.

Question 1b. Who made the decisions about which decontaminant to use in each place?

Response. For the Hart Building, EPA consulted with a variety of scientific resources, and selected the decontamination strategy in consultation with the Incident Commander. EPA has also consulted with scientific experts and provided extensive technical advice to the U.S. Postal Service in developing remediation methods and protocols for cleaning up postal facilities across the country. Overall, we have found that each site presents unique variables, and requires a site-specific cleanup plan that targets the most appropriate method for each contamination scenario.

Question 1c. Are bleach and the Sandia foam equally protective of human health?

Response. EPA does not have data on these two products with which to definitively compare their relative efficacy. However, EPA is conducting the "AOAC" sporicidal test on bleach, liquid chlorine dioxide, and Envirofoam to assure that they are effective at eliminating anthrax spores. Regardless of the question of efficacy, the Agency does not believe that either product, when used in accordance with the specified use directions, would cause any unreasonable adverse effects to humans or the environment.

RESPONSES BY HON. CHRISTINE TODD WHITMAN TO ADDITIONAL QUESTIONS FROM SENATOR SMITH

Question 1. In your oral testimony at the hearing you asked for the committee's help in two specific areas relative to remediation of buildings affected by bioterrorist attacks: limitation of liability and recovery of costs. Would you please explain in detail both your understand of current law in these two areas, and set forth with as much specificity as you are able, exactly what changes you seek?

Response. During the course of our response to anthrax contamination, EPA faced some issues that we had not previously encountered during the normal course of cleaning up hazardous waste sites under the Superfund program. I wanted to alert you to these important issues. The Agency has continued to work on finding ways to resolve these issues.

The first issue dealt with contractor reluctance to clean up anthrax under existing Comprehensive Environmental Response, Compensation and Liability Act (CERCLA) indemnification provisions. Given the many unknown factors associated with cleaning up substances resulting from acts of terrorism, some contractors were reluctant to perform anthrax cleanup work unless indemnification provided by EPA for potential Federal liability was also extended for potential state strict liability risks. Currently, EPA does not have the authority to indemnify contractors for strict liability to third parties under state law.

The second issue dealt with EPA's lack of authority to recover Agency cleanup costs associated with response to materials defined under CERCLA as pollutants and contaminants. Anthrax is considered a pollutant and contaminant under CERCLA. Under CERCLA, EPA only has the authority to recover cleanup costs for response to hazardous substances. It is unclear, however, whether broader liability under CERCLA would have a practical impact on the Agency's ability to accomplish the cleanup of anthrax contamination, because in many cases the contamination may be caused by an unknown third party. EPA would like to work with the committee to discuss options to address these issues.

Question 2. With respect to the remediation of the Hart Building, press reports state that although the fumigation procedure went well, the decision was made at the last minute to increase the exposure time from 12 hours to 20 hours. Would you please discuss this decision, and in particular outline all factors giving rise to the need to the increase in time, and state why you felt it appropriate to alter your plan at inception?

Response. Prior to the fumigation effort for the Daschle suite, EPA, working with chlorine dioxide industry representatives and the U.S. Postal Service, performed a number of tests at the Brentwood Postal facility on the effectiveness of chlorine dioxide in killing anthrax spores. These tests focused on the key variables which could influence the effectiveness of the product to kill spores. Those key factors include: concentration, contact time, temperature and humidity. Based on those tests, it is believed that a concentration of 750 ppm chlorine dioxide for 12 hours (or 9000 ppm-hrs) should be the objective to provide for the most effect kill. A decision was made during the operation to continue for a longer period of time in order to achieve the target concentration of 750 ppm of chlorine dioxide for 12 hours. There were difficulties in the first hours reaching the 750 parts per million (ppm) of chlorine dioxide concentration.

Question 3. Please discuss, with as much particularity as you are able, giving reference to all applicable statutes, regulations and case law, the basis for EPA's authority for having an onsite coordinator (OSC) for clean-up and/or remediation of a site of a biological hazard, including but not limited to sites which are contaminated by virtue of a terrorist attack.

Response. Biological hazards are pollutants or contaminants under the definition in the Comprehensive Environmental Response, Compensation and Liability Act (CERCLA) Section 101(33). Section 104(a) of CERCLA gives the President (who delegated the authority to EPA in Executive Order 12580) the authority to respond to releases or substantial threats of releases into the environment of any pollutant or contaminant which may present an imminent and substantial danger to the public health or welfare. 40 CFR 300.120, which is part of the National Oil and Hazardous Substances Contingency Plan established by Executive Order, sets out the general responsibilities of on-scene coordinators.

EPA also provides assistance to other Federal agencies engaged in emergency response activities. For example, under several Presidential Decision Directives, EPA provides support to the FBI during the crisis phase of terrorist events and helps the Federal Emergency Management Agency (FEMA) in the management of the consequences of a terrorist attack. When the President makes a disaster declaration under the Stafford Act, EPA also assists FEMA in responding, especially when the response involves hazardous materials.

Question 4. With respect to your election of chlorine dioxide gas instead of vaporized hydrogen peroxide, would you please explain, in specificity, the relative advantages and disadvantages of one over the other?

Response. Chlorine dioxide gas has more penetrating power than vaporized hydrogen peroxide. The nature and extent of anthrax contamination in the Daschle suite required a chemical that can penetrate into cracks, crevices and porous surfaces. Vaporized hydrogen peroxide works well only on hard, non-porous surfaces, so chlorine dioxide gas was the preferred alternative in this setting.

Question 5. Were there other options considered other than chlorine dioxide and vaporized hydrogen peroxide, such as ozone or other technologies?

Response. In consultation with a variety of internal and external scientific resources, EPA considered a number of options. The primary alternatives that are appropriate for office settings include High Efficiency Particulate Air (HEPA) filter vacuuming, chlorine dioxide gas, chlorine dioxide liquid and hypochlorite (bleach) solution.

Question 6. Is EPA working on the establishment of a process and protocol for seeking out new technologies for consideration to deal with any future contamination by biological weapons?

Response. EPA's Technology Innovation Office is leading an effort to collect and disseminate information about technologies to detect and decontaminate biological agents. We have established a web site "Technology for Biological Threats" http://EPATechBiT.org as a clearinghouse for information about these technologies and their vendors, and links to other resources pertaining to the detection and decontamination of biological agents. This website also helps vendors start the application process to have their antimicrobial pesticide product reviewed and registered in accordance with the Federal Insecticide, Fungicide and Rodenticide Act. We are operating a vendor helpline at (703) 390–0701 and an email address at EPATechBiT@ttemi.com to field inquiries from vendors of detection, decontamination, and measurement technologies. EPA's on-scene coordinators, emergency response personnel, and their contractors who are responding to incidents involving biological agents receive up to date information on new products and vendors collected by the hotline on a weekly basis.

EPA is also working closely with the Interagency Group on Terrorism's Technical Support Working Group (TSWG), jointly chaired by the Departments of State, Defense and Justice, to develop a formal process for selecting and approving new technologies for dealing with terrorism. The Department of Defense recently issued a Broad Agency Announcement for technologies that support the Federal Government's counterterrorism efforts, to help identify promising new approaches for decontamination, and detection of biological threats. In addition, EPA is working directly with TSWG to review promising new antimicrobial devices and detection technologies from vendors that have contacted EPA's Vendor Helpline. Much of the expertise to evaluate these innovative technical approaches resides in other agencies. TSWG is providing access to national experts to review and assess vendor claims.

Question 7. Is chlorine dioxide gas registered with your Agency as a sporicide?

Response. Yes, chlorine dioxide gas is registered with EPA as a sterilant to kill spores of bacillus subtilis and clostridium sporogenes. It has not however been registered specifically for bacillus anthracis spores. The Agency has approved its use under section 18 of the Federal Insecticide, Fungicide and Rodenticide Act for emergency use in anthrax decontamination plans that include thorough post-treatment sampling of decontamination areas to ensure effective decontamination.

Question 8. Would you please comment on the suggestion that chlorine dioxide has actually been occasionally unsuccessful in room decontamination in the pharmaceutical context?

Response. EPA is not aware of any such problems.

Question 9. You testified at the hearing, in response to one Senator's question, that Anthrax was detected at one spot in the HVAC system of the Hart Building. Please describe, in as much detail as possible, the efforts that were undertaken to eliminate any possibility of the presence of anthrax in other areas of the HVAC system, particularly un-tested surfaces.

Response. Hundreds of samples have been taken throughout the HVAC in an attempt to characterize the existence or location of anthrax. With the exception of the air handling unit which returns air from the Daschle suite, we have not found the presence of anthrax. This is consistent with the understanding that the anthrax in other locations was the result of the cross contamination of mail with the Daschle letter or foot tracking by personnel who moved between offices, rather than an airborne release.

Because of the potential that anthrax spores may exist undiscovered in the two air handling systems interconnected to the Daschle suite, it was determined that the most protective approach for public health was to clean up those air handling systems. This plan addresses both the air handling units and the connected ventilation ductwork on the return side of those units.

A number of cleanup options were considered including manual cleaning, liquid and foams, steam cleaning and fumigants. After consultation with HVAC experts and personnel with extensive knowledge on anthrax, it was determined that the most effective technology would be the application of chlorine dioxide gas. An attempt to fumigate the air handling systems connected to the Daschle suite was attempted over the weekend of December 14–16 but was eventually halted due to a number of mechanical difficulties, coupled with delays in achieving the optimal level of humidity. After working extensively to correct the problems which had arisen, the chlorine dioxide fumigation of the HVAC was implemented between the Christmas and New Year's holidays. This fumigation effort was successfully performed in that we were able to achieve our target goal of 9000 ppm-hours of chlorine dioxide. Subsequent sampling and analysis found no positive hits for anthrax and showed that a pervasive sterilizing effect had been achieved throughout the system.

Question 10. Were the protocols you developed for the remediation of Hart subject to scientific peer review? If so, would you please submit a summary of all reports generated in that process or processes?

Response. The peer review comments on the proposed fumigation of the entire Hart Building were submitted to the committee on December 14, 2001.

Question 11. What lessons have you learned in preparing for the remediation of the Hart Building that will guide you in similar future projects?

Response. The Agency has gained valuable experience during these clean-up activities. EPA, in conjunction with the interagency emergency response team and the U.S. Postal Services is in the process of identifying a comprehensive analysis of the lessons learned, which we will share with you shortly. The Agency is committed to working with you and your colleagues to ensure that should face a similar situation in the future, we will have full integrated the lessons we've learned from this experience.

Among the preliminary lessons, EPA recognizes the need for enhanced and improved medical monitoring and health and safety programs for our On-Scene Coordinators (OSCs) to ensure they are adequately protected when they are responding to bio-terrorism events, such as the anthrax release at the Hart building. Medical monitoring could include pre-response screening, response treatment (w/anaphylaxis antibiotics) and post-response followup and medical care for our OSCs, contractors and other responders. EPA also recognizes the need for more stockpiles of personal protective gear and key response equipment, including state-of-the-art field screening and bio-agent detection equipment. The Agency also needs to examine its alternative biological analytical capabilities and availability of trained technicians to determine if they are sufficient in the event additional or larger-scale anthrax/biological incidents were to occur.

Question 12. Please describe, in as much particularity as possible, how the EPA came to be involved with the anthrax contamination of all private entities (such as AMI in Florida), and provide references to all statutory and decisional law you relied upon in both commencing and in terminating your involvement.

Response. EPA tailors its involvement at individual sites according to the complexity of the problem, the urgency of the need, and the needs of the owner/operator of a facility. EPA has authority under the Comprehensive Environmental Response, Compensation and Liability Act (CERCLA) and an Executive Order establishing the National Oil and Hazardous Substances Contingency Plan (NCP), to clean up contamination within a building if the Agency believes that there is the possibility that the contamination may leave the building and create a release to the environment. As part of its efforts, EPA may deploy Federal resources to do monitoring, sampling, risk assessment, safety and health analysis, clean up, disposal, and other response requirements.

The EPA may defer to the owner/operator of a facility (whether private, Federal or State/local) to carry out a response to an incident involving hazardous substances and oil, if the Agency determines that they have appropriate response capability. In these instances, the EPA provides oversight and technical support. EPA has the authority to take over such responses, if necessary, in order to protect public health and the environment.

EPA became involved at the AMI building in Boca Raton, FL, at the request of the Florida Department of Health as well as the Palm Beach County Department of Heath. EPA has conducted assessment activities with the objective of insuring that the building did not pose an imminent and substantial threat to the surrounding community. The initial request also included EPA participating in a collaborative effort with the Health Agencies in determining how the building could be decontaminated. This action was conducted under the National Contingency Plan (NCP), developed by Exectutive Order, which authorizes the Agency to provide a Federal On Scene Coordinator for releases or potential releases of pollutants or contaminates that may present such threats. At this time, EPA has concluded that the AMI building constitutes effective containment for the potential anthrax release, and doesn't present a threat of release to the environment. The cleanup of contamination inside the building is the responsibility of the property owner and operator. This position is also consistent with the general approach of EPA's CERCLA response program.

EPA was not requested to provide assistance to the agencies (FBI, State and local law enforcement and health agencies) responding to anthrax releases that occurred at several sites in New York City. All assessment or cleanup operations were conducted by the building owners and operators. EPA has been in communication with the responding agencies to offer technical assistance, if needed.

In Connecticut, EPA has responded as part of an interagency investigative team. No contamination has been detected in private buildings.

Question 13. Please describe, in as much particularity as you are able, the nature and scope of the EPA's involvement at each anthrax contaminated building (public or private) in terms of its contamination with anthrax.

Response. At the American Media, Inc. (AMI) Building in Boca Raton, FL, EPA conducted comprehensive anthrax sampling to characterize extent of contamination. The sampling was initiated at the request of the State Health Department, and conducted with assistance from the Centers for Disease Control and Prevention (CDC), the Agency for Toxic Substances and Disease Registry, the U.S. Army Medical Research Institute of Infectious Diseases, U.S. Coast Guard Strike Team, the National Institute of Occupational Safety and Health, Palm Beach County Health Department, and EPA contractors. EPA sampling efforts included vacuum samples from soft surfaces such as carpet; wipe samples from hard surfaces such as desks; and air samples. All samples were analyzed for the presence of anthrax. EPA is currently providing technical assistance to the owner in developing and carrying out a strategy for decontaminating the building, including providing information on cleanup technologies, sampling protocols, and post-cleanup sampling. EPA will continue to work closely with AMI, along with other health agencies and local authorities, as cleanup of the AMI building proceeds.

At five Florida postal facilities and the Capitol Hill complex (Hart, Dirksen, Russell, Longworth, Cannon, Rayburn and Ford Office Buildings, as well as the House side of the Capitol), EPA with the assistance of CDC conducted sampling activities to confirm and determine the extent of contamination and cleanup activities. Sampling efforts included vacuum samples from soft surfaces such as carpet; wipe samples from hard surfaces such as desks; and air samples. All samples were analyzed for the presence of anthrax. In each of these cases, EPA's activities were conducted under the overall management of an Incident Commander provided by the owner/operator of the facility, which was the U.S. Postal Service (USPS), for the Florida Postal Facility sites, and the U.S. Capitol Police, for the Capitol Hill building sites. EPA also conducted cleanup activities at the five Florida postal facilities and the Capitol Hill Complex. The cleanup at the Florida postal facilities included decontamination using a bleach solution. Cleanup activities at the Capitol Hill complex included construction of isolation barriers in some office suites, fumigation using chlorine dioxide gas, decontamination by hand of hot spots using chlorine dioxide liquid and foams, and HEPA vacuuming. After cleanup activities are complete, EPA will conduct environmental sampling to ensure the effectiveness of the cleanup actions.

In addition to the postal facilities in Florida, EPA has provided technical assistance to the USPS for sampling and decontaminating other postal facilities throughout the country. The USPS is managing the responses under either its own authorities or as an executive agency under the NCP.

At other federally owned sites in the DC area, EPA is providing technical assistance for sampling and decontaminating the facility to the other agencies involved. In addition, EPA is working with GSA to provide the assessment and mitigation services needed for the responses.

Question 14. If any other agencies have to your knowledge had any involvement either directly or indirectly with the contamination of the privately owned facilities, please identify each such agency and state the nature and extent of that agency's involvement.

Response. The Federal Bureau of Investigation (FBI) was involved in anthrax incidents at the ABC, CBS and NBC media building in New York City, and the American Media, Inc. (AMI) building in Boca Raton, FL. The FBI's role has been to identify crime scene evidence and criminal intent.

The Centers for Disease Control and Prevention (CDC), the National Institute for Occupational Safety and Health (NIOSH) and the Agency for Toxic Substance and Disease Registry (ATSDR) provided technical assistance at postal facilities and the AMI building. CDC assisted with environmental sampling of the media buildings in New York. CDC and ATSDR also conducted environmental sampling at the USPS facility in Wallingford, CT.

The U.S. Coast Guard—Strike Teams provided technical assistance at the American Media, Inc building in Boca Raton, FL.

The U.S. Army Medical Research Institute of Infectious Diseases (USAMRIID) provided technical assistance at the American Media, Inc building in Boca Raton, FL.

The U.S. Army Corps of Engineers (USACE) provided technical assistance to the USPS for the decontamination of the Wallingford, CT, facility.

Question 15. Who do you believe is financially responsible for remedial costs related to anthrax contamination of a privately owned facility?

Response. Generally, private building owners are responsible for hiring qualified contractors to conduct sampling and perform whatever decontamination is necessary. Depending on the insurance coverage in force, there may be insurance money to defray the costs of cleanup. When asked, EPA provides an On Scene Coordinator to provide technical assistance. Under the National Contingency Plan, EPA has the authority to perform work if the situation exceeds the capabilities of the owner or state and local responders.

Question 16. With respect to the preceding question, please explain with as much particularity as you are able, the reasons upon which you base your answer, making reference to all applicable statutory and decisional law. Under what statute or under what authority will the remediation of these contaminated facilities be conducted?

Response. Clean-up of these facilities is authorized under the Comprehensive Environmental Response, Compensation and Liability Act (CERCLA). Anthrax is defined as a pollutant or contaminant rather than a hazardous substance, under the Section 101(14) and (33) of CERCLA. Under Section 104(a) of CERCLA, the President has the authority to respond to releases or substantial threats of releases of any pollutant or contaminant which may present an imminent and substantial danger to the public health or welfare. This authority has been delegated in Executive Order 12580. While EPA has broad authority to respond, the Agency's resources are limited and choices must be made about which responses to undertake using our limited resources.

CERCLA does not specify which parties are responsible for response costs associated with clean ups of pollutants or contaminants. Section 107(a) of CERCLA, which specifies which persons are responsible for response costs, deals only with costs incurred in response to releases or threatened releases of hazardous substances and not pollutants or contaminants.

Question 17. Who will establish and enforce the cleanup standard to be used at these contaminated facilities?

Response. There are no existing standards for cleaning up anthrax that has been deliberately released into a workplace setting. EPA has worked very closely with the Centers for Disease Control and Prevention, the Occupational Safety and Health Administration, the National Institute for Occupational Safety and Health, the Agency for Toxic Substance and Disease Registry and other experts, to evaluate the unique characteristics of each contaminated site, to recommend appropriate cleanup goals, and to evaluate the effectiveness of clean-up activities. The clean-up goal recommended to the Chairman of the Capitol Police Board for the Capitol Hill Complex was "zero growth," which means that there is no viable anthrax detected in any post-cleanup samples. Clean-up activities continued until this goal was met. We will continue to work with these Federal experts, and also with private owners, to recommend appropriate goals and evaluate effectiveness of remediation at other contaminated sites.

Question 18. What funding source was used to pay for your efforts, including but not limited to testing, of all non-public buildings?

Response. EPA's activities to address anthrax contamination were conducted using our emergency response authority. Therefore these activities were funded from our Superfund account.

Question 19. Please discuss any financial programs that exist that could be used to partially or fully underwrite the cost to remediate private facilities, such as Brownfields grants, technical grants to local health departments and EPA Regional Strategic Geographic Initiative discretionary grants?

Response. EPA does not have any financial programs to underwrite the cost of cleaning up private facilities. EPA's Local Governments Reimbursement program provides Federal funds to local governments for costs related to temporary emergency measures conducted in response to releases or threatened releases of hazardous substances. The program serves as a "safety net" to provide supplemental funding to local governments that do not have funds available to pay for these response actions. The Comprehensive Environmental Response, Compensation and Liability Act (CERCLA) specifically limits reimbursement to $25,000 per single response. CERCLA also specifies that only a small percentage of the Superfund budget can be used for local government reimbursement. The $25,000 cap, plus the limited availability of funds for the program, may not allow EPA to reimburse local governments for all response costs that may qualify.

Brownfields cooperative agreements could not help underwrite the cost of remediating private facilities contaminated with anthrax. Brownfields cleanup actions would not be timely enough. They are limited to non-time critical removals (i.e., non-emergency activities) and they must include several procedural steps, including public notice and comment. Also, private parties are not eligible for direct grants of brownfields funds. EPA awards cooperative agreements to eligible local governments, States, and Indian tribes to establish a "revolving" cleanup fund. The cooperative agreement recipient may issue loans to eligible public, private or non-profit borrowers to be used as cleanup funds for prospective projects.

It would not be appropriate to utilize Regional Strategic Geographic Initiative discretionary grants for private site clean up, as these funds are set aside for addressing unique regional/geographic issues rather than individual sites, working with the local community and all stakeholders.

Question 20. Have the testing protocols employed by the EPA at the Hart Building differed in any respect from the testing protocols employed by the EPA at other similarly contaminated facilities, and if so, please explain all reasons for this difference, and please explain the nature of the difference.

Response. Testing protocols may differ, depending on size and type of the potentially affected areas (e.g., a large open mailroom or office space with cubicles), how the contamination was delivered (e.g., spores in an envelope or by contact with a contaminated surface), how contamination could be dispersed (e.g., by "tracking" from a contaminated area or through an air handling system), and other industrial hygiene issues specific to the site. In each case, we attempted to recreate the likely path of the contamination source and sample along that path. If we found indications of contamination, we went back and sampled those areas more comprehensively, and designed a remediation plan to meet the specific characteristics of that site. Our knowledge of the effectiveness of different sampling and analytical methods has been growing day by day, and we are using what we learn to develop new approaches for future improvements to our response capability.

Question 21. What will be the EPA's future role in the remediation of private facilities contaminated by biological warfare agents, especially in light of Presidential Decision Directive 62?

Response. EPA is currently re-evaluating various response authorities, including those under the Comprehensive Environmental Response, Compensation and Liability Act, the National Contingency Plan, and Presidential Decision Directives 39 and 62, to determine the appropriate EPA role. We are also consulting with the Office of Homeland Security on this issue.

STATEMENT OF PATRICK J. MEEHAN, M.D., DIRECTOR, DIVISION OF EMERGENCY AND ENVIRONMENTAL HEALTH SERVICES, NATIONAL CENTER FOR ENVIRONMENTAL HEALTH, CENTERS FOR DISEASE CONTROL AND PREVENTION, DEPARTMENT OF HEALTH AND HUMAN SERVICES

Good morning, Mr. Chairman and members of the subcommittee. I am Dr. Patrick Meehan, Director, Division of Emergency and Environmental Health Services, Centers for Disease Control and Prevention (CDC), Department of Health and Human Services (HHS). Thank you for the invitation to discuss CDC's and HHS's Agency for Toxic Substances and Disease Registry's (ATSDR) role in supporting the Environmental Protection Agency (EPA) in remediating anthrax-contaminated workplaces. My division includes CDC's National Pharmaceutical Stockpile (NPS) as well as coordination of emergency preparedness and response activities under the Federal Response Plan.

Today, I will update you on the intentional release of anthrax and the number of exposed and affected persons, as well as summarize CDC and ATSDR's efforts to identify exposure, prevent anthrax disease, and monitor the health of those known to be exposed. I will also discuss CDC and ATSDR's collaboration with the EPA to assist in remediating contaminated buildings and protecting the health of workers in those buildings.

I would like to begin by emphasizing the importance of remediating all anthrax-contaminated worksites. CDC and ATSDR have worked, and will continue to work diligently along with EPA and our Federal, State, and local public health partners to help achieve this goal. Every worker in the United States deserves a safe and healthy workplace. In the past 2 months, terrorists have used anthrax spores to disrupt, displace, and even infect American workers. One phase of the fight against terrorism is to remediate contaminated workplaces and protect the health and safety of American workers who need to return to their jobs. We must also protect those workers whose job it is to investigate and clean these work places. These are the people who have been on the front lines of this battle, and they deserve our help and support.

As you are aware, many facilities in communities around the country have received anthrax threat letters. Most were received as empty envelopes; some have contained powdery substances. However, in some cases, actual anthrax exposures have occurred. These cases have been identified in Florida, New Jersey, New York, Washington, DC, and Connecticut. This is the first bioterrorism-related anthrax attack in the United States, and the public health ramifications of this attack continue to evolve. In collaboration with State and local health and law enforcement officials, CDC, ATSDR, and the Federal Bureau of Investigation (FBI) are continuing to conduct health investigations related to anthrax exposures. During this heightened surveillance, cases of illness that may reasonably resemble symptoms of anthrax have been thoroughly reviewed. The public health and medical communities continue to be on a heightened level of disease monitoring to ensure that any potential exposure is recognized and that appropriate medical evaluations are given. This is an example of the disease monitoring system in action, and that system is working.

PUBLIC HEALTH LEADERSHIP

The Department of Health and Human Services' (DHHS) anti-bioterrorism efforts are focused on improving the Nation's public health surveillance network to quickly detect and identify the biological agent that has been released; strengthening the capacities for medical response, especially at the local level; expanding the stockpile of pharmaceuticals for use when needed; expanding research on disease agents that might be released, rapid methods for identifying biological agents, and improved treatments and vaccines; and regulating the shipment of hazardous biological agents or toxins.

As the Nation's disease prevention and control agency, it is CDC's responsibility on behalf of DHHS to provide national leadership in the public health and medical communities in a concerted effort to detect, diagnose, respond to, and prevent illnesses, including those that occur as a result of a deliberate release of biological agents. This task is an integral part of CDC's overall mission to monitor and protect the health of the U.S. population.

REMEDIATION SUPPORT ACTIVITIES

Since the intentional release of anthrax spores, one of the areas on which CDC and ATSDR have focused is the identification and cleanup of contaminated facilities. We have refined methods for environmental sampling to assess whether anthrax contamination had occurred; in buildings that has meant sampling of air and surfaces. CDC and ATSDR have issued recommendations on how to conduct environmental sampling and how laboratories should analyze those samples. We also recommended environmental sampling strategies to characterize the extent of exposure and to guide cleanup. We issued recommendations to protect first responders, investigators, and cleanup personnel. As buildings were identified as contaminated, we provided technical input to EPA and others tasked with cleanup to determine where remediation was necessary. These recommendations have been widely disseminated to Federal, State and local health and environmental agencies, and are available at CDC's bioterrorism website (*http://www.bt.cdc.gov*).

EPA has devised strategies for remediation and has gained much experience through its activities to date. Disease experts at CDC are developing strategies to prevent the spread of disease during and after bioterrorist attacks. Although there are some data on chemical disinfectants in the scientific literature, there are no historical data that indicate the best way to eliminate spores from an office building, or to disinfect a sorting machine. The ability of a disinfectant to kill an anthrax spore is dependent upon time of contact and concentration and is mitigated by the amount and composition of material through which it must penetrate to get to the spore. For many of the clean-up methods being used to kill anthrax spores, we will not know their effectiveness until we go through the process. EPA understands this and has sought help from a variety of sources, including CDC and ATSDR, to ensure that the appropriate indicators are used and that post-sampling strategies are adequate.

With regard to the effectiveness of cleaning, even our most exhaustive sampling strategies will not identify every spore. It is unlikely that any cleaning strategy will kill every spore. However, the EPA should be able to clean and re-test to the point where we all are comfortable that spores have been killed or removed from surfaces where human contact is likely to occur. A range of sampling methods and strategies should be used to ensure the safety of building occupants.

In heavily contaminated areas, such as Senator Daschle's suite and the Brentwood postal facility, fumigation is being proposed as the method of clean-up. The use of fumigants is a potential hazard for clean-up workers, those in areas adjacent to the buildings, and those that must re-occupy the building. A fumigant that is effective at killing spores is, of necessity, a highly toxic agent. The protection of workers during the fumigation process is a matter of good industrial hygiene. EPA, CDC, and ATSDR are working together to ensure remediation workers are protected during the fumigation processes. EPA works with local public health agencies to ensure that people in the area but outside of the building being fumigated are notified and kept at a safe distance.

With regard to the safety of those who will re-occupy the building, it is important to determine both that the area is clear of the fumigant and that there is no health risk. Again, CDC, ATSDR, and the Occupational Safety and Health Administration (OSHA) have developed exposure limits for fumigants, and detection methods are available to determine when any residual fumigant is well below established limits. After buildings are cleaned and post-cleaning environmental sampling has been conducted, CDC and ATSDR are committed to providing technical input to the incident command and other experts to determine whether the building is ready for re-occupancy.

CHALLENGES

CDC has been addressing issues of detection, epidemiologic investigation, diagnostics, and enhanced infrastructure and communications as part of its overall bioterrorism preparedness strategies. Based on Federal, State, and local response in the weeks following the events of September 11 and on recent training experiences, CDC has learned valuable lessons and identified gaps that exist in bioterrorism preparedness and response at Federal, State, and local levels. CDC will continue to work with partners to address challenges such as improving coordination among other Federal agencies during a response and understanding the necessary relationship needed between conducting a criminal investigation versus an epidemiologic case investigation. These issues, as well as overall preparedness planning at Federal, State, and local levels, require additional action to ensure that the Nation is fully prepared to respond to acts of biological and chemical terrorism.

CONCLUSION

In conclusion, CDC and ATSDR are committed to working with other Federal agencies and partners as well as State and local public health departments to ensure the health and medical care of our citizens. We are committed to continuing our partnership with EPA to ensure that the best public health information is coupled with the best ideas for how to remediate contaminated facilities. We need to improve sampling methods and equipment. We must learn from this experience and continue to assist the EPA in determining the best ways to remediate different types of workplace environments having different amounts of anthrax contamination.

Thank you very much for your attention. I will be happy to answer any questions you may have.

RESPONSES BY TRACY MEEHAN TO ADDITIONAL QUESTIONS FROM SENATOR JEFFORDS

Question 1. Although the risk of secondary aerosolization is very slight, there is still a risk. Therefore, what precautions are you proposing? What are your thoughts on post-exposure vaccination?

Response. On December 18, 2001, the Department of Health and Human Services released a statement on their recommendations for post-exposure vaccination and continued prophylaxis. The introduction of this release reads as follows:

Many of those who were exposed to inhalational anthrax in the recent mail attacks are presently concluding their 60-day course or preventive antibiotic treatment. Some of these persons especially those who may have been exposed to very high levels of anthrax spores, may wish to take additional precautions. The Department of Health and Human Services (HHS) is providing two additional options beyond the 60-day antibiotic course, for those who may wish to pursue them: an extended course of antibiotics, and investigational post-exposure treatment with anthrax vaccine.

HHS will make anthrax vaccine available to those who were exposed to inhalational anthrax, who have concluded their antibiotic treatment and who wish to receive the vaccine as an investigational product. The vaccine is being made available in this investigational mode, under an investigational new drug application (IND) at the option of the individual, in recognition of the limited nature of the data now available concerning inhalation anthrax treatment and the factors underlying development of the disease, as well as uncertainty concerning the extent of exposure to spores that some persons may have received in the recent anthrax incidents. The decision to use this vaccine is at the discretion of the individual, in consultation with his or her physician.

The complete document can be accessed at: *http://www.hhs.gov/news/press/2001pres/20011218.html.*

Question 2. Can you describe the protocol used to assess potential health risks of a chemical or biological attack?

Response. Since there are a number of agents which can produce a biological or chemical attack and a variety of exposure routes, each episode is investigated to determine the public health risks based on the scientific circumstances of the incident. The investigation process includes an epidemiological investigation to determine the source and mode of transmission and define the at-risk population; a laboratory investigation to define the agent; an environmental assessment to detect any potential exposure to a chemical or biological agent, including how long the exposure lasted; an evaluation to detect any evidence of clinical disease; and a followup to determine any potential long-term health risks.

Question 3. I understand that you have received test results of various remediation technologies, for example about ECASOL. What are your impressions of those test results?

Response. We evaluate remediation technologies based on past history with that technology, published studies, experience in cases of biologic contamination, and the effect of the technology on the contaminated media. Some technologies, such as ECASOL are new, experimental, or have only been used by the military and will require further review. Others, such as sodium hypochlorite and Sandia Foam show some evidence of effectiveness against anthrax spores, and we recommend their use with appropriate followup sampling on a case-by-case basis.

Question 4. Much of the level of risk associated with anthrax exposure depends on the degree to which anthrax is treatable with antibiotics. How does your protocol account for individuals who are allergic to, are pregnant or have compromised immune systems and cannot take the antibiotics used to treat anthrax without putting themselves at serious health risk or inflicting serious harm on unborn children? Would you agree that the potential risks associated with anthrax contamination are greater for an individual who is in some way limited in the antibiotics available to them?

Response. Antibiotic regimens are available that include alternatives for persons in the described groups. The following MMWRs contain interim recommendations for alternative prophylaxis:

Notice to Readers: Update: Interim Recommendations for Antimicrobial Prophylaxis for Children and Breastfeeding Mothers and Treatment of Children with Anthrax. Vol 50, No 45;1014 11/16/2001 *http://www.cdc.gov/mmwr/PDF/wk/mm5045.pdf*

Notice to Readers: Updated Recommendations for Antimicrobial Prophylaxis Among Asymptomatic Pregnant Women After Exposure to Bacillus anthracis Vol 50, No 43;960 11/02/2001 http://www.cdc.gov/mmwr/PDF/wk/mm5043.pdf

Question 5. Can you identify for me how you determined which individuals should be taking 60 days of antibiotics? Even for those that tested negative for anthrax, you have suggested this regiment, in effect prescribing Cipro as a preventative measure? How may this protocol change in the future?

Response. Sixty days of prophylactic antibiotics were recommended for persons who had significant exposure to powder containing anthrax spores or were in environments in which they were at risk of exposure or where there were cases of inhalational disease. The nasal swab tests which were conducted in some settings were not done for diagnostic purposes, and a negative nasal swab did not mean that a person was not exposed to anthrax. The test was used mainly for epidemiologic purposes to determine an area where people were likely exposed and for forensic purposes. Decisions regarding which groups of individuals should receive prophylaxis were made in collaboration with Federal, State, and local health officials and partners in the areas impacted over the course of the investigation.

Many of those who were exposed to inhalational anthrax in the recent mail attacks are presently concluding their 60-day course of preventive antibiotic treatment. Some of these persons, especially those who may have been exposed to very high levels of anthrax spores, may wish to take additional precautions. The Department of Health and Human Services (HHS) is providing two additional options beyond the 60-day antibiotic course, for those who may wish to pursue them: an extended course of antibiotics, and investigational post-exposure treatment with anthrax vaccine. HHS will make anthrax vaccine available to those who were exposed to inhalational anthrax, who have concluded their antibiotic treatment and who wish to receive the vaccine as an investigational product.

The complete document can be accessed at (*http://www.hhs.gov/news/press/2001pres/20011218.html.*)

RESPONSE BY TRACY MEEHAN TO ADDITIONAL QUESTION FROM SENATOR CORZINE

Question. As we clean up the anthrax mailed in October, we should be thinking about how we prevent future incidents in the future and mitigate the effects of any incidents that do occur. One of the ideas that has been discussed in this regard is vaccination. Do you think that the anthrax vaccine should be made available an on option to postal workers and others who may be at high risk for future anthrax exposure?

Response. Supplements to the December 15, 2000 ACIP recommendations that concern the use of anthrax vaccine are currently under review at CDC and HHS. These include persons at potential repeated exposure to anthrax. Examples include laboratory and decontamination workers. Postal workers have not been demonstrated to be at similar risk.

RESPONSES BY TRACY MEEHAN TO ADDITIONAL QUESTIONS FROM SENATOR SMITH

Question 1. Please state the date of your first involvement with the decontamination and remediation of:

(1) The Hart Building.
(2) The AMI Building.
(3) The NBC Offices.
(4) The New York Post Offices.
(5) The West Trenton, NJ Postal Facility.
(6) The Stevens' home.
(7) The New York City Office of Gov. Pataki.
(8) The Boca Raton Postal Facility.
(9) The Brentwood Postal Facility.
(10) The Ford Office Building.
(11) The Ottilie Lundgren home.

Response. In all of the sites involved in the recent anthrax episode, discussions regarding site decontamination and remediation were woven into the overall public health investigation and response. Therefore, rather than specify dates of first involvement with decontamination and remediation, we are providing dates CDC initiated onsite investigations related to anthrax in each of these locations.

Florida, October 4.
New York City, October 12.
New Jersey, October 18.
Washington, DC, October 15.
Connecticut, November 20.

Regarding the specific locations listed, it should be noted that there was never any anthrax detected in the environmental samples taken from the Stevens and Lundgren homes. Therefore, CDC is not aware of any issues regarding decontamination and remediation in these locations. In addition, CDC was not involved in the investigation of the New York city office of Governor Pataki. This sampling and cleanup activity was conducted by the New York State Department of Health; CDC did provide some support for specimen confirmation. In the other settings, CDC served in a consultative role to the incident command, the Environmental Protection Agency (EPA), and to state and local health and environmental authorities.

CDC works closely with the EPA to evaluate human health risk from environmental anthrax contamination and the adequacy of decontamination. As noted from Governor Whitman's previous testimony before Congress, decontamination is an EPA responsibility:

Under the provisions of PDD 62, signed by President Clinton in 1998, the EPA is assigned lead responsibility for cleaning up buildings and other sites contaminated by chemical or biological agents as a result of an act of terrorism. This responsibility draws on our decades of experience in cleaning up sites contaminated by toxins through prior practices or accidents.

Since the intentional release of anthrax spores, CDC and ATSDR focus has been on the identification and characterization of anthrax contamination in facilities either where cases have been identified or which have been associated with the investigation (such as "downstream" mail facilities), in addition to the clinical, laboratory and epidemiologic investigation of the cases of anthrax infection. We have refined methods for environmental sampling to assess whether anthrax contamination had occurred; in buildings that has meant sampling of air and surfaces. CDC and ATSDR have issued recommendations on how to conduct environmental sampling and how laboratories should analyze those samples.

We also recommended environmental sampling strategies to characterize the extent of exposure and to guide cleanup. We issued recommendations to protect first responders, investigators, and cleanup personnel. As buildings were identified as contaminated, we provided technical input to EPA and others tasked with cleanup to determine where remediation was necessary. EPA, CDC, and ATSDR are working together to ensure remediation workers are protected during the fumigation processes. After buildings are cleaned and post-cleaning environmental sampling has been conducted, CDC and ATSDR are committed to providing technical input to the incident command and other experts to determine whether the building is ready for re-occupancy.

Question 2. Another witness at the hearing testified that the standard for certification of safety to re-occupy the Hart Building, and impliedly any contaminated building, following remediation, is that 100 percent of all tests for spores is returned negative—that no anthrax spores are detected. Please explain your position in this regard, with as much specificity as you are able.

Response. Although CDC does not "certify" in a regulatory sense that a building is safe, we agree with the stated position. After remediation, when we are asked to consult, as in the Hart Building, we will review the pre- and post-remediation sampling strategy and results. We will consider the Hart Building safe for re-occupancy when appropriate remediation has occurred and when a rigorous sampling strategy shows no detectable spores. It must be noted, however, that sampling cannot ever evaluate every surface in a building and we can never say that every spore has been killed.

Question 3. With respect to your answer to the preceding question, please provide copies of all standards or protocols that have been developed.

Response. Our position regarding the interpretation of post-remediation samples represents what CDC believes to be good industrial hygiene practice, but is not explicitly stated in a published document on anthrax remediation. As noted in our answer to question No. 1 above, we have published protocols for sampling, for the laboratory analysis of samples and for the protection of personnel doing sampling and clean up. They can be found on our web site: *www.bt.cdc.gov.*

Protection of personnel doing sampling and clean up . . .
http://www.bt.cdc.gov/DocumentsApp/Anthrax/Protective/Protective.asp
Sampling . . .
http://www.bt.cdc.gov/DocumentsApp/Anthrax/11132001/final42.asp
Laboratory Analysis . . .
http://www.bt.cdc.gov/Agent/Anthrax/LevelAProtocol/Anthracis20010417.pdf

---

STATEMENT OF IVAN C.A. WALKS, M.D., CHIEF HEALTH OFFICER, DISTRICT OF COLUMBIA AND DIRECTOR OF HEALTH

Good morning, Chairman Jeffords and distinguished members of the Committee on the Environment and Public Works. I am Dr. Ivan C.A. Walks, Chief Health Officer for the District of Columbia and Director of the Department of Health. With me today is Theodore J. Gordon, Chief Operating Officer of the Department of Health (DOH), and key staff members involved with the remediation of biologically and chemically contaminated buildings. We appreciate the opportunity to testify and commend you for convening this Hearing because the discussion here this morning further complements our effort to illuminate the issues regarding environmental exposures to contaminants in the District of Columbia.

This hearing also enhances our effort to continuously inform the community and involve them in decisions or procedures designed to address their concerns. As I mentioned in the Hearing on Spring Valley before the House of Representatives in July 2001, we cannot overemphasize the importance of an ongoing interaction between the District of Columbia Government and the members of the community. There can be no substitute for an informed community. That theme has been and will continue to be a guiding light for our efforts in every community in the District of Columbia, and in any other effort to prevent disease, dysfunction and premature death.

Allow me now to turn to the purpose of this hearing, i.e., the process that the District of Columbia Government is guided by in remediating biologically and chemically contaminated buildings and its progress and successes to date. My testimony will also cover the challenges that confront the District and the rest of the country, the new technologies available, and our next steps.

The DOH's Environmental Health Administration is charged with the mission of protecting human health via the prevention and control of environmentally related diseases, the prevention of environmental degradation, and the promotion and preservation of the ecological system and physical environment of the District of Columbia. When carrying out this charge, it is imperative that we follow a process that is structured, but at the same time flexible enough to allow for input from the various stakeholders. In this regard, and particularly with regard to time-critical remediations, our process is similar to that described by the U.S. Environmental Protection Agency's "Superfund Community Involvement Handbook."

DISTRICT'S PROCESS OF REMEDIATION

The first step toward remediation that we take is to identify/define the problem. Regarding biological contamination, this step involves both the identification of contaminated regions of a building and all the possible pathways by which contamination can move beyond the contaminated zone to other locations within the building.

We then begin to explore the various remediation options. Each option is evaluated with regard to its technical effectiveness, practical feasibility, and the unintended health and ecological risks to remediation workers and the adjacent community. In other words, we perform an environmental risk assessment identifying issues and the problems or risks associated with each option.

In conducting an environmental risk assessment, several things are considered. First, we must be confident that we will achieve a successful outcome. Also with regard to each option, we also have to consider cost, exposure to the government, community hardship (emotional and physical), and length of time for the cleanup. We continue to monitor and re-evaluate throughout the planning and implementation stages of the process.

From this, a prime option is then identified. We also develop a secondary or "fall back" option so that we do not have to restart from the beginning if the prime option is not selected.

Once we are almost certain that we have considered all pertinent factors, we then prepare to take a plan of action to the affected stakeholders for input and "buy in." We have learned a long time ago that there is no such thing as a successful plan, if the community has not had the opportunity to participate in it. Again, a big reason for our success in the Spring Valley community had to do with the inclusion of that community in our remediation strategy. We have had several meetings in the community briefing its residents on our findings and process for remediating. In addition, Mayor Anthony Williams assembled an independent group, the Spring Valley Scientific Advisory Panel, which includes seven specialists in the fields of epidemiology, toxicology and environmental health, as well as two representatives from the Spring Valley community.

The DOH has had significant experience in remediating biologically and chemically contaminated buildings in the District of Columbia. Within the most recent 18 months we have experienced Legionella contamination in a correctional facility, a public school, and in a health care facility. We have had significant fungal contamination of private homes and a public high school following a flood this past summer. In one community, private homes and a District building were affected by a petroleum spill. Our successes are largely attributable to how well we communicate with the affected parties. Of course, we have a highly skilled and professional group of scientists and engineers who perform the technical risk assessment and remediation steps discussed above. However, I continue to stress the importance of communication as a key ingredient in any successful remediation plan.

CHALLENGES CONFRONTING THE DISTRICT OF COLUMBIA AND THE COUNTRY

The particular challenge confronting the DOH in the District of Columbia and all health departments across the Nation regarding biological decontamination of buildings is that the these remediations necessarily must take place in a context of emerging science. We are all traveling steep learning curves with regard to the technical and medical facts. When we use toxic chemicals to kill biological agents, the scope of that learning curve includes stakeholders both within and adjacent to the affected building.

In this regard, we wish to recommend one fundamental public health principle: until we learn whether a clinically significant minimum microbial contamination level exists, we should only declare a building to have been decontaminated when all test samples achieve “no detection” levels. With regard to community exposure to toxic chemicals we should continue to maintain substantial margins of safety with regard to exposures to people in the adjacent communities.

NEXT STEPS

As we proceed to climb these learning curves, we need to share information with other State and local health agencies. Such information will include biological sampling protocols, dosing, measuring, critical bio-load levels and most of all, effectiveness data. We should expect the emergence of new chemical decontamination methods, rapid measuring technologies, and biological detection methods. Knowledge of their efficacies and protocols should be widely shared within the public health community.

Thank you for this opportunity to come before you to discuss this issue. We are happy to answer any questions you may have.

STATEMENT OF MIKE GROSSER, U.S. MARINE CORPS, TECHNICAL DIRECTOR, PROGRAM MANAGER, NUCLEAR, BIOLOGICAL AND CHEMICAL DEFENSE SYSTEMS, MARINE CORPS SYSTEMS COMMAND

Mr. Chairman and members of the committee, I am Mr. Mike Grosser, the Technical Director for the Program Manager, Nuclear, Biological and Chemical Defense Systems, Marine Corps Systems Command, Quantico, VA. I am pleased to appear before you today to discuss several decontamination technologies that the Marine Corps and the Joint Chemical and Biological Defense community have been developing and supporting. I am responsible to the Program Manager for the oversight of these programs and I have knowledge of the origin, progress and current status of each.

The Marine Corps has pursued these technologies as possible solutions to the requirement for an environmentally benign, patient-friendly and effective personnel and equipment decontaminant. We did not set out to identify a specific decontaminant for anthrax-contaminated buildings. The technologies that I will talk about are by and large still in research and development. They have been, and in fact still are considered as candidates for the Joint Service Family of Decontaminating Systems Program and may be designated as more appropriate for use by the first or secondary responders, that is, a municipal firefighter or a unit such as the

Marine Corps Chemical-Biological Incident Response Force (CBIRF), than the traditional warfighter. While it is possible that one or two of them may be made available quickly, each has some facet that still requires funding, research, testing or evaluation. I will describe four decontamination technologies.

The first technology, Electrochemical Activated Solution, or ECASOL, was developed in 1972 in Russia to control oil well biofilms. It is now used commercially in Russia, Japan, South Africa and the United Kingdom where it is used for home drinking water purification units (300,000 units sold) and as a hospital biocide such as patient decontamination, surface decontamination, surgical device sterilization, wastewater treatment and is also used for reducing pathogens in food processing operations (e.g. meat and poultry). ECASOL was used to purify drinking water in Rwanda during the refugee crisis in 1994–1995.

ECASOL is a colorless, odorless aqueous solution made onsite using point-of-use electrolysis of diluted brine. The brine is exposed to a mild electrical charge as it passes through a patented Flow-through Electrolytic Module (FEM), a 10" by 1" diameter tubular device that converts the brine into a stream of reactive oxidants. A key benefit of the ECASOL technology is that the oxidant composition can be precisely controlled over a wide pH range. pH is a measure of the acidity or alkalinity of a solution. A pH level of 1 is acidic, and a pH level of 14 is an alkaline. For personnel decontamination, skin contact requires a near neutral pH. At neutral pH (pH 7) the primary oxidant in ECASOL is the metastable compound hypochlorous acid. This acid, though safe to skin, eyes and wounds (pH 7), is an effective biocidal agent. The primary military personnel decontaminant for medical application is 0.5 percent HTH (bleach) which has a pH of 12, is irritating to the skin and not safe for eyes or wounds.

The Marine Corps began testing ECASOL in 1998 to assess safety, efficacy and the potential to scale-up field units for use with first or secondary response personnel. Tests were designed to compare ECASOL's efficacy versus 0.5 percent (5,000 ppm) bleach at destroying biological and chemical agents.

Some chemical testing has been conducted but the results were not as promising as those obtained during biological agent tests.

Before 1998 the largest ECASOL unit was an 80 FEM (400 Gallon per hour) unit used in a poultry processing plant. Based on the above test results the Marine Corps built a 600-gallon per hour prototype generator to evaluate the potential for use by first response units in personnel showers. All volume generation targets and solution parameters were met or exceeded during field trials of that unit.

Although ECASOL is generated onsite at the point of use, shelf-life or storage characteristics were examined. ECASOL solutions stored in sealed containers for 7 weeks were found to perform almost as effectively as freshly generated solutions. Solution parameters of pH, free chlorine and oxidation-reduction potential showed some deterioration (<10 percent), although overall performance was maintained. Again, while this information is important, the intent of the technology is to produce the decontaminant on site.

Further evaluation is required to identify maximum and minimum effective concentration ranges, effective pH range, efficacy against Toxic Industrial Chemicals and Toxic Industrial Materials (TIC/TIM), evaluation as an aerosol (fog), and potential for decontaminating waste runoff.

Materials and components required to generate the ECASOL are salt and water (or brine, seawater), electricity and a device containing FEMs.

ECASOL effluent is environmentally benign and can be drained into a municipality's sewer system (demonstrated in Atlanta, GA and Camp Lejeune, NC).

To summarize, ECASOL is a highly effective biocidal agent. It has a major advantage over 0.5 percent bleach because it has a neutral pH (7) and is safe for eyes, wounds and skin whereas bleach has a pH of 12, irritates skin and is not safe for eyes or wounds. Although the technology works with aqueous solutions ranging from saturated brine (for producing chlorine) or just plain water (for water purification) in dilute solutions (as examined here) it is safe yet effective. The technology is flexible and has been demonstrated in large scale (600 gallons per hour) as well as small scale (5 gallons per hour) applications.

The ECASOL device developed for testing by the Marine Corps could be utilized to conduct the test for room/building decontamination proof of principle. To produce additional prototypes would require purchase of some custom made long lead items and manufacturing. Three additional prototypes could be functional and delivered in approximately 120 days. In the interim, the existing device is capable of producing 600 gallons of product per hour. A comprehensive test plan has already been developed for additional efficacy testing (chemical and biological) that will include additional live agent testing.

The second promising technology is electrostatic decontamination (ESD) currently under development at the University of Missouri in Columbia, MO. This research and development program was started in 1998.

ESD is an electrostatically charged mist containing a proprietary photosensitizer that is sprayed onto a contaminated surface, victim or a wound. The photosensitizer consists of a hydrogen peroxide base (1–2 percent), a proprietary additive, and a surfactant. The photosensitizer is then illuminated with a pulsed ultraviolet (UV) light source that activates the photosensitizer destroying all biological agents present. System efficacy against chemical agents is unknown as no tests have been completed at this time. The photosensitizer mist is harmless and will not cause damage or injury to humans or the environment. The pulsed UV light wavelength is used for only 4 to 60 seconds and is not harmful to humans. Eye protection can be provided by regular glasses or by simply closing your eyes. The system operates in ambient conditions from temperatures ranging from freezing to 120° F and provides open-air sterilization.

Testing revealed the following destruction times:

Photosensitizer + Pulsed UV light:

Anthrax spores, 75 seconds;

E. coli bacteria, 75 seconds;

Salmonella, 75 seconds;

Water borne virus simulants, 75 seconds.

Photosensitizer only—No Pulsed UV light:

Anthrax spores, 8 minutes;

E. coli bacteria, 8 minutes;

Salmonella, 8 minutes;

Water borne virus simulants, 8 minutes.

These results are based on using twice the density of spores required by NATO standards.

The ESD system is comprised of four major elements:

1. Proprietary photosensitizer-hydrogen peroxide solution,
2. Spray applicator,
3. Ultraviolet light source, and
4. Water

All of these elements are commercial-off-the-shelf (COTS) items with the exception of the proprietary photosensitizer. The shelf life of the photosensitizer is 1 to 3 years depending on the purity of the hydrogen peroxide used. Application of the mist shows coverage of 100 m^2/10 liters in 9 minutes. I'd like to note that it has not been developed or evaluated as a room or ductwork decontaminant, but rather as a surface decontaminant; we believe that ESD can be misted into enclosed spaces or ductwork to effectively neutralize biological agents.

This developmental effort would require some minor modification of COTS applicators, and testing to ensure proper procedures are in place to maximize agent neutralization in a building/ductwork environment. The effort could be completed in 6–8 months if the appropriate test facilities are made available.

The third technology is a nanoparticle regime that includes materials with particle sizes ranging between 1–100 nanometers (1 nanometer = 10^{-9} meters). Nanoparticles of metal oxides exhibit extraordinary abilities to react with and thereby destroy highly toxic substances and chemical warfare agents. Kansas State University (KSU) and their commercial adjunct firm, Nanoscale Materials, Incorporated (NMI), have been active since 1995 in developing metal oxide nanoparticles and defining their applications with regard to destructive adsorption.

Recently, it was also found that special formulations of these nanoparticles are active against biological warfare agents such as spores of *Bacillus globigii*, which is a simulant of anthrax. With respect to biological agents, nanoparticles have a positive charge that enables them to attach to negatively charged bacteria cells or spores. Once attached to the bacteria or spore the nanoparticle penetrates the cell walls of bacteria destroying the nucleus. For thick-coated protein cells of spores, addition of chlorine as a stabilized free radical to the nanoparticle formulation enhances their ability to penetrate these cells.

Since August of this year Marine Corps Systems Command has aggressively pursued this technology for a wide range of decontamination applications. This project is focused on developing novel dry powder decontamination technologies capable of neutralizing chemical and biological warfare agents. With appropriate funding this technology could be available for use as a biological decontaminant as soon as calendar year 2003.

Sandia National Laboratory (SNL) has developed the fourth technology.

This decontaminant (designated DF–100) is a non-toxic, non-corrosive aqueous foam with enhanced physical stability for the rapid mitigation and decontamination of chemical and biological warfare agents and toxic industrial materials. The foam formulation is based on a surfactant system to solubolize contaminants and increase reaction rates with nucleophilic reagents and mild oxidizing agents. The formulation includes water-soluble polymers to enhance the physical stability of the foam. Preliminary test results demonstrate very effective decontamination of chemical and biological threat agent simulants on contaminated surfaces and in solution. Testing also indicates that the formulation may be effective as a general decontaminant on a variety of toxic industrial materials. This decontamination technology offers the following benefits: (1) a single decontaminant solution for both chemical and biological threats (2) rapidly deployable (3) minimal operational and logistics impacts.

Studies conducted on the DF–100 decontaminant to date include chemical agent decontamination efficacy (post-decon contact and off-gas vapor hazards), reaction rates, detector compatibility, toxicity, materials compatibility and biological simulant decon efficacy. Biological simulants tested to date include anthrax and smallpox simulants.

Chemical testing revealed that DF–100 destroyed 99–100 percent of G, V and H class agents in 10–60 minutes. Biological testing revealed that DF–100 was effective in reducing biological simulants to a safe level. Of particular interest, in a 10E6 challenge (1M spores) using Bacillus globigii (Anthrax Simulants), SNL Foam achieved a 6-log reduction (reduced to 1 spore or less) within 15 minutes. Other simulants tested included smallpox and *E. Coli* MS2 with similar results.

Material characteristics include a pH of 9.8 and a liquid to foam expansion of 15:1.

Currently two companies are licensed to manufacture and produce DF–100. These companies also manufacture or are licensed to sell application systems capable of dispensing DF–100. These application systems range in size from man-portable (back pack system) to truck mounted. Included in these application systems is the Marine Corps Compressed Air Foam System (CAFSM), a HMMWV mounted fire fighting system.

Discussion with industry indicates that manufacturing facilities are capable of producing up to 20,000 gallons per day of DF–100. Production/delivery capabilities for application systems range from 1000 per month for small systems to 20 per month for large systems.

Decontamination demonstrations at Dugway Proving Ground and Fort Leonard Wood have shown that DF–100 may be applied with currently fielded decontamination systems or dual use systems i.e. firefighting systems, pressure washers.

Preliminary evaluations and studies conducted on SNL DF–100 under the Joint Service Family of Decontamination Systems program were designed against tactical operational requirements. SNL DF–100 has not been evaluated for room or interior decontamination under the JSFDS program to date.

In conclusion, Mr. Chairman, I want to thank the committee for inviting me to present this information. This is a vitally important issue to the Marine Corps and to our Homeland Defense. The Marine Corps and the Joint Chemical-Biological Defense Program continue to conduct research, development and acquisition of these and other technologies with the sole intent of providing Marines and other service members with the very best capability. I will be happy to address any questions at this time.

Statement of Les C. Vinney, President and CEO, STERIS Corporation

Mr. Chairman and members of the committee, good morning. My name is Les Vinney. I am president and chief executive officer of STERIS Corporation. I thank you for your invitation and welcome the opportunity to address this critically important issue given the unprecedented challenge that we face as a Nation.

I am accompanied this morning by Dr. Peter Burke, STERIS vice president and chief technology officer, and Mr. Gerry Reis, STERIS senior vice president, Corporate Administration. Also joining me is Ms. Karla Perri, senior environmental consultant of Versar, Inc.

STERIS Corporation has $800 million in revenues and is a New York Stock Exchange publicly traded company. STERIS technologies are used every day in environments where the highest levels of sterility are required. Healthcare professionals in virtually all hospitals across the United States, and scientists and researchers in the pharmaceutical industry—including the Fortune 50 pharmaceutical companies—use STERIS products to sterilize and decontaminate items, from surgical instruments to their equipment and facilities. These technologies help ensure positive outcomes of such critical activities as the production of antibiotics, the development of vaccines, and the safety of sensitive medical devices and implants for human beings.

In its simplest form, the primary business focus of STERIS is to develop and produce formulations that prevent infection and contamination, and the delivery systems to enable their most efficient use. When properly utilized, these technologies can provide safe and effective remediation of contaminated materials in whatever form they may take, including entire rooms and their various contents. These technologies can also be put in place to prevent recontamination and assure ongoing safety, just as is their purpose in the industries we currently serve.

In light of recent events in our country, we welcome the opportunity to offer our experience to help prevent infection and contamination, and to clean and restore biologically contaminated facilities for normal use. Our persistence in offering our technologies for these applications is driven by the belief that our technologies can help to optimize and improve the safety of the current remediation efforts, both in their application and potential residual effects.

Toward that end, we have joined with Versar, Inc., a leader in providing counterterrorism, environmental, architectural, engineering and related services. Together, STERIS and Versar offer a broad array of contamination risk assessment and remediation services.

Mr. Chairman, we firmly believe that methods now in use in healthcare and scientific settings can effectively decontaminate facilities infected with anthrax. The reason that you have not previously seen us before your committee is that the large majority of STERIS products are traditionally used in hospitals and by pharmaceutical companies. As such, we normally have had our technologies and processes accepted for use under the purview of the Food and Drug Administration.

While many of our formulations have been registered for specific uses with the Environmental Protection Agency, our decontamination processes have not previously been registered for specific applications, such as mail and building decontamination, of the kind our Nation is now addressing.

Since the initial anthrax contamination events, we have had numerous meetings with officials on Capitol Hill and in various Federal agencies to discuss the possible uses for our products and services. While our past experience gives us very high confidence in the effectiveness of our technologies, we strongly endorse the regulatory requirements to test and validate a product technology prior to allowing its use in specific treatment applications.

In that regard, we have been seeking the opportunity to demonstrate the efficacy of our product technologies to meet various remediation needs—and allow people to safely return to their work environment. We hope a bridge can be created across regulatory jurisdictions to enable the more rapid application of these existing capabilities to meet emergency decontamination needs.

We are now working closely with the EPA in the attempt to secure the necessary approvals to permit the use of these available applications. We are also in advanced discussions with the Department of Justice on a potential demonstration project, which would serve to validate the effectiveness of these technologies in decontaminating anthrax infected facilities.

In recent years, hazardous materials decontamination efforts have largely focused on remediation of contaminated water and soil. Buildings contaminated with anthrax present an unprecedented challenge.

Effective remediation requires multiple technologies to deal with both microbial and biochemical contaminants.

The healthcare and pharmaceutical industries have dealt with microbial control challenges for many years. As a result, highly sophisticated prevention and treatment methodologies have been developed within these industries. While older technologies such as formaldehyde and chlorine dioxide have, in fact, been used in these industries, newer technologies, such as vapor hydrogen peroxide and the combination of hydrogen peroxide and peracetic acid sporicidal compounds, have been developed. These emerging technologies have displaced the earlier technologies because they offer reduced toxicity, limited corrosiveness, minimal residual effects, and easier application.

A facility contaminated by highly aerosolized anthrax spores, which have been distributed to remote areas due to cross-contamination during mail delivery or through ventilation systems, involves a unique and severe challenge. While these conditions present a different environment than our more standard applications, we believe our technologies can be applied to the remediation and elimination of contaminants in this type of setting, as well.

To accomplish proper remediation, a carefully planned process similar to the Hazard Analysis and Critical Control Point approach would be used, just as is currently done in establishing the preventive process for healthcare and scientific requirements. In an appendix attached to my written testimony we have presented a detailed plan for systematic biological remediation of a given facility or area.

For any remediation effort, STERIS working with Versar recommends a series of steps to render a contaminated area safe for use. These include mapping the extent of contamination, reviewing the area and its contents, decontaminating using a combination of technologies and methods, confirming effectiveness and documentation.

It is also important to note that the length of any remediation process will depend on the scope of the project—including the level of contamination—and size of the building. All of the proper biological indicators and others tests must be completed before employees can be allowed to return to a building.

Mr. Chairman and members of the committee, in our professional view there is no single silver bullet for treating chemical or biological contamination. This remediation requires the selective use of multiple technologies, not reliance on a single treatment type. This approach should result in the least damage to items within contaminated facilities, assure that each surface and material is treated with the agent best suited to its individual needs and provide the highest level of decontamination.

In closing, we believe a coordinated effort is needed among the appropriate government, academic, military and private industry officials. This coordinated approach will permit the identification, validation and utilization of the safest and most effective technologies currently available. Careful development of the proper protocols for this remediation process is critical to a successful outcome. What we must achieve is the restoration and maintenance of safe working environments for all Americans. STERIS stands ready to help.

Thank you for the opportunity to appear before you today. I would be happy to answer any questions you may have.

APPENDIX A

STERIS Corporation Overview

STERIS Corporation is a leading provider of infection prevention, contamination prevention, and microbial reduction products, services, and technologies to healthcare, scientific, research, food, and industrial customers throughout the world. Founded in 1987, and expanded with a series of acquisitions of companies with over 100 years of service, STERIS has been at the forefront of meeting customers' needs to prevent infection and contamination, contain costs, and improve efficiencies. STERIS products can be found wherever there is a need to ensure the highest levels of sterility.

Headquartered in Mentor, Ohio, the Company has 4,500 employees, with production and manufacturing operations in 14 States plus Puerto Rico, Canada, Finland and Germany. The Company has sales offices located in 17 countries. STERIS has annual sales of over $800 million, and its stock is traded on the New York Stock Exchange under the symbol STE.

STERIS customers include more than 5000 hospitals, Fortune 50 pharmaceutical companies, and many leading medical device manufacturers. The Company's broad array of infection and contamination prevention products and services are used every day by healthcare professionals, scientists and researchers to ensure that materials and surfaces are free of contamination and safe for human contact. STERIS technologies are also used to decontaminate critical environments such as clean rooms, isolators, and research work areas.

STERIS professionals are committed to understanding the needs of each individual customer and customizing the application of the Company's technologies to ensure positive outcomes of such critical activities as the production and manufacture of medicines to prevent and cure disease, to eliminate the risk of infection during surgical procedures, and to ensure that sensitive medical devices and implants are safe for use on human beings.

The Company is committed to the development of new technologies as well as the discovery of new applications of existing technologies, to serve the infection and contamination needs of its customers. The Company's core technologies and services include:

- High and low temperature sterilization systems utilizing steam, ethylene oxide, vaporized hydrogen peroxide, and paracetic acid based technologies.
- Contract sterilization services provided through a network of 16 facilities in North America offering gamma irradiation, electron beam and ethylene oxide sterilization technologies.
- Surface disinfectants and liquid cold sterilants formulated to disinfect and sterilize hard surfaces.

- Personnel hand wash and rinse products that are used to keep hands free of bacteria.
- Surgical support products and services that enable healthcare professionals to provide the highest levels of patient care.
- Automated washing/decontamination systems and related detergent and cleaning chemistries.
- Facility planning and design services.
- Contamination risk assessment and remediation services.
- Education, training, installation and repair services.

APPENDIX B

DETAILED BIOLOGICAL REMEDIATION PLAN

EXECUTIVE SUMMARY

Let us briefly consider the technologies that are available and our objectives in their use. These antimicrobial technologies should be rapidly effective at killing bacterial spores, which of all microorganisms are accepted as the most difficult to kill. Further, they should have minimum safety hazards, not damage the room or its important contents, and if possible be widely used and accepted for decontamination.

First, certain room contents including rugs, drapes, personal items, and electronic equipment may need to be removed and decontaminated separately from the room. STERIS recommends that these can be batch sterilized by widely used methods including ethylene oxide or irradiation. It may be also prudent to consider the overall cost of remediating these items compared to the alternative of removing, appropriately disposing and replacing them.

Technologies available to decontaminate rooms may be divided into two categories: liquid and gaseous.

A variety of liquid and foam-based technologies are available. In general, most routinely used disinfectants in households and hospitals demonstrate relatively slow or indeed no activity against bacterial spores. For example, high concentrations of chlorine solutions (like household bleach) are not recommended due to limited activity against spores and damage to surfaces. STERIS recommends the use of EPA-registered sporicidal products that are currently used for this purpose in high-risk or regulated areas, which have past rigorous, standardized tests and have demonstrated material compatibility.

Overall, liquids or foams are excellent for small surface application, but are difficult to ensure coverage and effectiveness over larger areas (including walls and ceilings). They also require significant time for application and cleanup, and will not be practical for certain surfaces, including electrical equipment.

Gaseous or vapor technologies are recommended for rooms. The most widely used are formaldehyde and Vaporized Hydrogen Peroxide (VHP). Formaldehyde is less used today due to variable efficacy and significant health and safety concerns. VHP has been widely used and accepted as a safe alternative. This dry process has been used for over 10 years in the pharmaceutical industry for room decontamination and has been validated for use in a government facility for anthrax decontamination. A simple, mobile VHP system generates, supplies, controls and neutralizes the dry vapor into a given area in one stand-alone process. A low concentration of vapor is required to rapidly kill spores, but is also very compatible with surfaces, including electronics and painted surfaces. This technology is one of the safest and an equally effective method for room decontamination.

DETAILED ANALYSIS

STERIS recommends that HACCP (Hazard Analysis and Critical Control Point) principles should be applied, since in our opinion no single intervention to this situation will be adequate to reduce the risk 100 percent. The basis of HACCP is to identify and to conduct a hazard (or risk) analysis, identify critical control points and introduce controls (or interventions) at these points to reduce contamination from *Bacillus anthracis*. It is further clear that no single technology is applicable or capable of complete decontamination in every area, but combining technologies and products that have been widely used, registered and accepted for similar applications in other environments should be adopted. A logical series of steps can be taken to maximize the decontamination process:

- Buildings should be sealed and contamination mapped. High and low risk areas should be identified and interventions (either single or multiple) conducted to reduce infection risks associated with each area.

- A combination of methods employed for decontamination:
  - HEPA vacuuming or surface liquid treatment (this in many cases may be sufficient, depending on the level and scope of contamination)
  - Boxing up of absorptive materials in heavily contaminated rooms and sterilizing by irradiation, ethylene oxide or terminal destruction.
  - Preparation of area for decontamination and any pretreatment with liquid sporicidal agents.

Products used should have demonstrated (and registered) broad-spectrum antimicrobial activity on a surface as well as material compatibility.

- Room fumigation with sporicidal, registered and material compatible process. This may be alone suitable as a preventative measure in room with low or suspected no contamination where surface decontamination of room contents may be sufficient depending on the determined risk.
- Verification of process effectiveness by process monitoring and documentation
- Retesting for contamination following decontamination to confirm effectiveness.

In general, the remediation plans that are under discussion for anthrax-contaminated buildings do adopt HACCP principles, identifying the overall problem and recommending potential methods of remediation. However, the plan appears to critically rely on chlorine dioxide (ClO_2) gas as the primary disinfecting/sporicidal agent to decontaminate the building, as well as manual treatment with some foams and liquids, but relying in particularly on chlorine dioxide and concentrated bleach solutions. A number of alternative registered products that have been widely used for similar applications do not appear to have been considered for remediation of biologically contaminated buildings. A review of the remediation plan and products that could be used are discussed below.

It is important to note that bacterial spores, such as *Bacillus anthracis* spores, are traditionally considered the hardest of all microorganisms to kill. These spores are significantly more resistant than normal bacteria, viruses and fungi, and are difficult to eradicate using standard disinfection or decontamination methods. Therefore, in cases of contamination with anthrax spores, decontamination methods are required to show rapid and consistent sporicidal activity, but also compatibility with the surfaces being treated. Although a variety of simple microbiological methods may be used to indicate the possible effectiveness of a given product against bacterial spores, a specific registration is required in the United States. Any liquid, vapor or gas product that is registered with the EPA has shown effectiveness relative to a rigorous, standardized test, namely the AOAC International Sporicidal method. EPA registered and widely used sporicidal products should be considered first for decontamination against anthrax spores.

Overall no single method will be effective for all contaminated areas. In some cases, certain room contents may not be compatible with, may not be adequately decontaminated or may even inhibit the effectiveness of the decontamination method. These items may include rugs, drapes, personal items, electronic equipment and paper, depending on the decontamination method used. It is recommended that these items have specific treatment plans to assure sporicidal effects. In some instances treatment in place with certain gaseous products is appropriate, while external treatment of other items should be employed. Batch sterilization of isolated items can be performed by widely used methods including ethylene oxide or irradiation, and returned to the room. Alternatively, following decontamination certain items may be destroyed by incineration. It may be also prudent to consider the overall cost of remediating these items compared to the alternative of removing, appropriately destroying, disposing and replacing them.

STERIS offers more than 28 years of sterilization experience and 16 sites throughout North America for irradiation and ETO sterilization. These facilities have processed more than 60 million cubic feet of product in the last 12 months, including medical supplies, pharmaceuticals, food containers, spices and cosmetics.

Irradiation is the process of exposing a product or material to ionizing radiation. Ionizing radiation is energy that exists in the form of waves and is defined by its wavelength. As the wavelength of energy gets shorter, the energy increases. Radiation destroys microorganisms by breaking chemical bonds in biologically important molecules such as DNA, and by creating free radicals and reactive molecules, which chemically attack the microorganism. Irradiation is not the same as radioactive. Many consumer products are sanitized, sterilized or modified by irradiation of the materials. Irradiation methods, their antimicrobial efficacy and applications are widely accepted and used for contract sterilization of wrapped and/or packaged materials and products, including medical devices and foods.

Ethylene oxide (ETO) is a colorless gas, which is used for the low temperature sterilization. Developed in the 1940's and 1950's, ETO is the primary gas used in hospitals to sterilize reusable items (e.g. medical devices that contain plastics) that cannot tolerate high sterilization temperatures. In addition, ETO sterilization is used for contract sterilization of medical, dental or veterinary devices that are delivered sterile to a consumer which are sensitive to steam sterilization or that contain materials incompatible with irradiation sterilization. The properties and broad-spectrum antimicrobial activity of ETO have been well described in the literature.

Technologies available to decontaminate potentially biological contaminated rooms, enclosed areas, HVAC ductwork, fixed and mobile equipment, and general hard surfaces may be divided into two categories: liquid and gaseous.

Liquid based technologies include a variety of products, which include liquids and foams. In general, most routinely used disinfectants in households and hospitals demonstrate relatively slow or no activity against bacterial spores. Products that are generally not effective include phenols and quaternary ammonium compound-based products. Sodium hypochlorite solutions (commonly referred to as 'bleach' or 'chlorine') can be effective but the following points need to be taken into consideration. At high concentrations, bleach will demonstrate some activity against spores; however, it requires long contact times, for example, purified spores placed directly into freshly prepared 10 percent bleach for 15–20 minutes will give an average 3-log reduction of spores. The effectiveness of bleach is dramatically reduced by interfering surfaces and organic soils, which also interact with the available chlorine. Furthermore, to our knowledge bleach is not a registered sporicide with the U.S. EPA. A further concern, which is familiar to all of us, is compatibility with room materials and surfaces; bleach, like other chlorine-based products can be damaging and even destructive to a variety of surfaces. Bleach can be effective over extended exposure times but only on clean, compatible surfaces.

A variety of other alternative liquid or foam formulations can also be recommended and maybe more applicable. These include oxidizing-agent based formulations, including liquid hydrogen peroxide, peracetic acid, chlorine dioxide or combinations thereof. We propose that any of these products, with demonstrated activity against a wide range of microorganisms, including bacterial spores, demonstrated material compatibility, reasonable safety and worker health profile, and, if possible, experience of use outside of a laboratory setting can be used for decontamination of anthrax. An example of an EPA-registered sporicidal product is SPOR-KLENZ, which is a liquid, synergistic combination of hydrogen peroxide and peracetic acid, which is widely used and validated for use in the pharmaceutical industry for its rapid spore killing activity. A complete dossier of publications, pharmaceutical applications, case studies, safety and user references are available.

There are also registered chlorine dioxide-based products, but in general these may be more damaging on surfaces. Certain foam or nanoemulsions have also been recommended. In comparison, these products require significantly longer contact times, have not been widely used and should also pass the required rigorous antimicrobial testing and safety profile for EPA registration.

Liquid or foam based products do have some major limitations. The most obvious is ensuring correct application of the product over all contact surfaces, including walls, floors, ceilings and room contents for the required decontamination time. For example, these products are not practical for HVAC ductwork. Following decontamination, the product also needs to be removed and dried prior to normal use. Additionally, surface compatibility with liquid or foam-based products varies depending on the product. Of greatest concern is the use of 'wet' methods relative to electrical equipment (including phones and computers), as well as other sensitive surfaces. In general, these products are not used or reliable for large, uncontrolled surface areas.

Gas or vapor-based technologies can also be considered, which possess acceptable registered spore killing activity, material compatibility, and safety/worker health profile. A summary of the advantages and disadvantages of these methods is attached in Table 1. The most widely used methods for this purpose are formaldehyde and Vaporized Hydrogen Peroxide (which is referred to as VHP). Formaldehyde has been traditionally used for over 100 years, although less frequently today due to variable efficacy and significant health and safety concerns. Formaldehyde is extremely toxic and carcinogenic. Further it leaves a white residue on all surfaces following the decontamination process, which is toxic and needs to be adequately removed prior to occupancy. From an effectiveness point of view, decontamination is relatively uncontrolled and usually takes up to 36 hours for completion. Of greatest significance is the fact that these rooms need to be humidified before and during treatment.

For these reasons, VHP has been used as an effective alternative. The VHP process is a rapid, dry, controlled technology using a low concentration of hydrogen peroxide vapor. Unlike liquid hydrogen peroxide, VHP is rapidly sporicidal at low concentrations and has been widely used as a validated process for over 10 years for room and enclosure decontamination. For example, the process is routinely validated for decontamination of rooms and enclosures using bacterial spores, and in certain selected cases against anthrax spores, to confirm process effectiveness. A simple, mobile system generates, supplies, controls and removes VHP from a given environment in a one step process, which can be monitored, verified and documented. Being a 'dry' method, the process demonstrates excellent compatibility with a wide range of materials, including paint and electrical equipment like computers. The VHP process is the safest method available for vapor/gas decontamination; for example decontamination may proceed in a sealed room while personnel safely work in adjacent areas and no cleanup is required following the process. One disadvantage is that the presence of significant cellulosic-based materials in a given room may elongate the process time and multiple generators are required to do areas larger than 7500 ft^3. A new high capacity VHP delivery and control system has recently been developed by STERIS to be available as soon as possible for large-scale room decontamination. A complete dossier of publications, pharmaceutical applications, case studies, safety and user references are appended.

Other technologies that may also be reasonable alternatives to formaldehyde include chlorine dioxide gas, which has shown good promise in the laboratory setting. Chlorine dioxide gas is rapidly antimicrobial but has significant material compatibility concerns. It is undetermined whether this process has been registered with the EPA, apart from a special exemption for anthrax decontamination. Like formaldehyde, significant humidification of a given area is required for chlorine dioxide gas to be effective in a room, which needs to be kept in the dark to prevent breakdown. Five hundreds times the concentration of chlorine dioxide gas is required to be present and maintained to be sporicidal relative to vapor hydrogen peroxide over a longer contact period (8 hours vs. 4 hours). Chlorine dioxide gas has a higher level of safety risks associated with its use and can also leave a white residue that requires immediate clean-up following decontamination. These safety risks also apply to its production, transport and use, as the gas cannot be easily produced onsite. For all these reasons, chlorine dioxide gas has not widely used or accepted for this application. Attempts to apply a controlled delivery and removal process based on chlorine dioxide gas for the decontamination of cleanrooms was unsuccessful in actual pharmaceutical, controlled applications.

STERIS has presented a rational, detailed plan for decontaminating biologically contaminated areas and their contents to render them safe for human contact. This plan recommends the use of multiple technologies for this purpose and recommends the use of EPA-registered products, which have been widely used for many years and remain the safest, effective and most practical methods available for room decontamination.

Table 1.—Comparison of Room Decontamination Methods

Fogging/foaming	Formaldehyde	Gaseous chlorine dioxide	VHP
Variable coverage and distribution.	Variable coverage and distribution.	Depending on mode of delivery, more reliable distribution. Difficult to maintain in gaseous state; can condense.	Controlled delivery system for more reliable distribution. Kinetics of maintaining gaseous state is understood and important for process effectiveness.
Wet methods	Requires significant hydration for antimicrobial efficacy. Essentially wet process.	Requires significant hydration for antimicrobial efficacy. Essentially wet process.	Dry sterilization method
High concentrations required for rapid sporicidal activity.	High concentrations required for rapid sporicidal activity.	500ppm sporicidal over 8 hours (but needs to be kept in the dark).	1–2ppm sporicidal at 25° C. 1 log reduction every 1–2 minutes.
Difficult to control and deliver over large surface areas and ensure residence time for horizontal surfaces.	Significant risks and difficultly providing to a large area. Overall better coverage than fogging or foaming.	Significant risks and difficultly providing to a large area. Overall better coverage than fogging or foaming.	Controlled delivery contacts all surfaces. New system available for large area fumigation
Difficult to validate	Difficult to validate	Validation possible. Can be biologically verified.	Validation and documentation routinely conducted. Can be parametrically, biologically and chemically verified.

Table 1.—Comparison of Room Decontamination Methods

Fogging/foaming	Formaldehyde	Gaseous chlorine dioxide	VHP
Concerns over material compatibility; extent dependant on contact time and antimicrobial agent/formulation.	Can be damaging to surfaces.	Significantly damaging to a variety of surfaces, even after single exposures. Concerns already noted in cleanroom applications.	Broad range material compatibility
Not safe on electrical equipment.	Not safe on electrical equipment.	Not safe on electrical equipment.	Safe on electrical equipment
Efficacy inhibited by presence of absorbing materials.	Stable, difficult to remove ...	Efficacy inhibited by presence of absorbing materials.	Efficacy inhibited by presence of absorbing materials
Occupational risks significant, dependant on antimicrobial used.	Significant occupational and safety risks.	Occupational risks significant, but can be minimized.	Occupational risks minimal. Safest for environment and personnel health
Extended contact times and clean-up required.	Extended contact times and clean-up required.	Extended contact times and possible clean up required. Chlorine residuals.	Most rapid method and room ready for use directly following cycle. No residuals.
Limited registration, depending on antimicrobial..	Limited registration, traditional use.	Unknown registration situation with process.	Sterilant used in the process registered with the EPA
Variable efficacy depending on the product.	Variable efficacy	Broad spectrum antimicrobial activity.	Broad-spectrum antimicrobial activity, including independent testing against and validation with B. anthracis.

RESPONSES BY LES C. VINNEY TO ADDITIONAL QUESTIONS FROM SENATOR JEFFORDS

Question 1. Have you approached EPA about your technologies for decontamination? If so, what type of response did you receive?

Response. STERIS officials have had numerous meetings with EPA officials to discuss our technologies for decontamination—and we have provided all of the documentation requested to validate our technologies. Among those we have met with are Marianne Horinko, Assistant Administrator for Solid Waste and Emergency Response, Claudia McMurry, Chief of Staff to Deputy Administrator Linda Fisher, Carlton Kempter, Senior Advisor, Antimicrobials Division, Office of Pesticide Programs, and Rich Ruppert, Site Coordinator for the Hart Building Remediation.

While many of our products and formulations are registered for specific uses with the EPA, our decontamination processes have not previously been registered for specific clean up applications, such as mail and building decontamination, stemming from the kind of bio-terrorism events our Nation is now addressing. The large majority of STERIS products are traditionally used in hospitals and by pharmaceutical companies. As such, we normally have had our technologies and processes accepted for use under the purview of the Food and Drug Administration. As such, the primary purpose of our meetings with the EPA has been to seek the necessary re-labeling of our products to allow their use for mail and building decontamination now taking place under their jurisdiction.

We are working closely with the EPA in an attempt to secure the necessary approvals to permit the use of these available applications—and move forward with their use. On December 12, 2001, we provided EPA with the formal request for specific exemptions clarifying our position for these STERIS technologies—vaporized hydrogen peroxide (VHP), Spor-Klenz® Peracetic Acid Sterilant and ethylene oxide sterilization systems. We have previously provided detailed background data to support this request.

On December 13, 2001, EPA officials contacted STERIS to ask for our assistance in the remediation of an EPA facility in Northern Virginia. We are hopeful, based on that request and subsequent conversations, we are close to receiving the relabelings that we have requested and will be able to begin work on mail and building decontamination in the very near future.

Question 2. Tell us about your VHP technology in terms of its ability to prevent recontamination.

Response. In high-risk areas like a mailroom, microbial decontamination can be accomplished by a routine regimen of disinfection to prevent possible cross-contamination during normal mail sorting. Unlike many other technologies, vaporized hydrogen peroxide (VHP) can be used without degradation to hard, non-porous surfaces on a routine basis for microbial destruction.

VHP technology provides rapid, low-temperature decontamination methods for any enclosed area that may be contaminated with microorganisms, including spore-forming bacteria. These systems are widely used to render surfaces and areas safe for contact. VHP systems are currently used as rapid, low temperature techniques for decontamination of producing filling lines, sterility testing environments, sealable enclosures, and various types of rooms in hundreds of installations, including pharmaceutical production, laboratory animal, research and biosafety laboratory facilities.

Further systems are also available to sterilize medical devices. In all of these situations, VHP technology renders contaminated surfaces safe for use or contact. There are no residuals or inhibitory chemicals remaining on a surface following the decontamination. For this reason, VHP is designed to render a surface safe for use, but when a surface is recontamined (*e.g.*, if a further anthrax-laced letter was opened in a given area), the process would have to be repeated in that area to render it safe for contact. For example, a pharmaceutical clean room used to manufacture an antibiotic, is routinely decontaminated using this technology on a weekly or even a daily basis to maintain the area as clean and sterile.

RESPONSES BY LES C. VINNEY TO ADDITIONAL QUESTIONS FROM SENATOR SMITH

Question 1. Please state, as directly as possible, and with as much particularity as you are able, any criticisms you have of the decisions made to date in the remediation process used at the Hart Building.

Response. EPA has never asked STERIS for a formal proposal to remediate the Hart Building. As a result, we have not had access to the facility since it has been contaminated—and, therefore, it is difficult to address the issue with any specifics. However, early on in the process we met with Secretary of the Senate Jeri Thomson to present our general suggestions on the Hart Building remediation. As a follow-up to that meeting, Ms. Thomson provided STERIS with a copy of Proposed Action Plan for Remediation of the Hart Senate Office Building (HSOB), Washington, DC (October 31, 2001) and asked for our analysis and comment. We provided our response on November 5, 2001, and I have included a complete copy of that document for your review (attached as Appendix A).

Question 2. In addition to any answer you give to the preceding question, please state directly your position as to whether VHP would have been a better choice as a fumigant over chlorine dioxide gas for use in the Hart Building.

Response. As I outlined in my testimony, the healthcare and pharmaceutical industries have dealt with microbial control challenges for many years. As a result, highly sophisticated prevention and treatment methodologies have been developed within these industries. While older technologies, including chlorine dioxide and formaldehyde have, in fact, been used in these industries, newer technologies, such as STERIS's VHP system, have been developed. The use of formaldehyde has decreased due to the fact that it is a known human carcinogen, and the use of chlorine dioxide has likewise decreased due to the very corrosive nature of the chemical. Therefore, emerging technologies like VHP have displaced the earlier technologies because they offer reduced toxicity, limited corrosiveness, minimal residual effects, and easier application. I would draw your attention to the chart previously submitted as the final page of my written testimony (attached as Appendix B)—a comparison of room decontamination methods, including fogging/foaming, formaldehyde, chlorine dioxide and VHP.

APPENDIX A

PRIVATE AND CONFIDENTIAL

NOVEMBER 5, 2001.

MS. JERI THOMSON, *Secretary of the Senate,*
U.S. Senate,
Washington, DC.

DEAR MS. THOMSON: Thank you for providing us with a draft copy of the Proposed Action Plan for Remediation of the Hart Senate Office Building (HSOB) Washington, DC. (October 31, 2001)—and permitting our analysis and response.

STERIS Corporation has specific questions on the reliance on chlorine dioxide (ClO_2) as the primary agent for decontamination of the HSOB. The plan is based on the assumption that ClO_2 will have sufficient penetration of items such as paper, rugs, and drapes. However, we believe there is reason to question this approach—both from a scientific perspective regarding the sporicidal effect of the treatment, as well as potential damage to materials in Senate offices, including fine art. Additionally, the proposed use of diluted bleach has important limitations with regard to sporicidal properties, which are not clearly expressed in the document.

Based on our questions, we have attached a detailed three-part response for your review:

- STERIS Corporation comments on the remediation plan (Tab A);
- Comparative tables of the attributes of chlorine dioxide versus vapor hydrogen peroxide and diluted liquid bleach relative to EPA-registered sporicidal products (Tab B); and
- An outline of factors in the planned remediation, with alternate considerations as a reference guide (Tab C).

STERIS Corporation is pleased to present these documents, which address the critical decontamination needs facing our government. We look forward to further discussions, and to offering our assistance in the remediation process.

Sincerely,

PETER A. BURKE, PH.D.,
Chief Technology Officer.

APPENDIX A

TAB A

COMMENTS ON THE PROPOSED REMEDIATION PLAN FOR THE HART SENATE BUILDING (HSOB)

The Hart Senate Office Building (HSOB) was contaminated on October 11, 2001; with reportedly weapons grade *Bacillus anthracis* spores in an envelope. A very concentrated spore population has left significant contamination in some areas of the building, while other sectors remain uncontaminated. This organism is projected by the Center for Disease Control (CDC) to have an Infectious Dose 50 percent (ID_{50}) of 6,000–10,000 spores and a potential for cutaneous anthrax with as little as 5–50 spores in animal trials. The ID_{50} reflects a normal, healthy population; however, any individuals that are in any way immunocompromised are more susceptible. Further, the persistence and potential germination/proliferation of spores under suitable conditions (*e.g.* damp air vents) could potentially contaminate further areas. Hence, a very conservative approach appears to be warranted.

STERIS recommends that HACCP (Hazard Analysis and Critical Control Point) principles should be applied, since in our opinion no single intervention to this situation will be adequate to reduce the risk 100 percent. The basis of HACCP is to identify and to conduct a hazard (or risk) analysis, identify critical control points and introduce controls (or interventions) at these points to reduce contamination from *Bacillus anthracis*. It is further clear that no single technology is applicable or capable of complete decontamination in every area, but combining technologies and products that have been widely used, registered and accepted for similar applications should be adopted.

The remediation plan of the HSOB, as discussed in the public press, does adopt HACCP principles, identifies the overall problem and recommends potential methods of remediation. However, the plan appears to critically rely on chlorine dioxide (ClO_2) fumigation as the primary disinfecting/sporicidal agent to decontaminate the building.

The major premise of the plan assumes that an overnight residence time utilizing at least 1000 ppm of chlorine dioxide will significantly reduce the microbial population without adversely affecting the furnishings, including artwork and personal effects, in offices. Based on our experience, STERIS would technically disagree with this assertion.

It is reasonable to ensure that decontamination of exposed surfaces will occur with ClO_2. ClO_2 has been widely used for disinfection of water, but is considered corrosive as a liquid surface disinfectant and particularly as a gaseous fumigant. ClO_2 is not widely used as a sporicidal fumigant, although it is used at low concentrations for odor control. Further, as an oxidizing agent, the degree of penetration into and efficacy on absorptive materials, including carpets, drapes, piles of papers, organic/inorganic soils and filing cabinets, is unknown and probably limited. It is also important to note that these materials just noted, due to their active absorption and breakdown of ClO_2, dramatically reduce the level of decontamination in a given area. This is true for any antimicrobial agent, including oxidizing agents.

It is recommended that the following points be considered for remediation:

- Consideration should be given to treating the rugs, drapes, personal items, electronic equipment and paper separately.

Rugs—Since the degree of ClO_2 penetration is unclear, perhaps direct treatment with a sporicidal product would be prudent to reduce hot spots of spores in the carpet pile. EPA registered rapid sporidical products should be considered first and used for this purpose. One class of products that should be considered are those based on peracetic acid/hydrogen peroxide combinations, in particular, due to the synergistic modes of action and unique attribute of peracetic acid to retain its antimicrobial activity in the presence of significant soil. These products could be directly applied prior to any gaseous treatment.

Drapes—It will be difficult to directly treat these items; however, they could be packaged and sent for offsite irradiation or ethylene oxide sterilization.

Consideration should be given to the overall cost of remediation of rugs and drapes, and to the alternative of removing, destroying and replacing these items.

Papers—The heavily contaminated offices should have the papers irradiated to allow disinfection of all hard surfaces without impedance. It is important to note that piles or files of paper may not be adequately penetrated or decontaminated using the proposed method.

- A further concern, if reported correctly, is the use of bleach (10 percent), which is slowly sporicidal and will be damaging for many surfaces at the specified contact times. It has been our experience, using purified spore preparations, that bleach diluted 1:10 shows a 2–3 log reduction of *Bacillus* spores after 15–20 minutes contact in a suspension test, which is under best case contact conditions. More rapid, efficacious and registered products for surface sporicidal activity are available and could be used.
- Chlorine dioxide, while an excellent sanitizing agent of water, is considered very reactive with organic materials, such as wood, as well as metals. The items described above, including paintwork, wood coverings, *etc.*, will most likely be damaged if fumigated with ClO_2. Significant residuals will also remain on surfaces following fumigation and may require lengthy aeration times or post-fumigation clean-up.

It is important to note that fumigation with ClO_2 requires significant hydration of all surfaces in order to be effective, which can be further damaging to certain surfaces due to the risk of condensation (for example, it would not be applicable with computers or other electronics), elongates the overall cycle time and is difficult to ensure with large room volumes. For all these reasons, ClO_2 is not widely used for room decontamination.

The most widely used antimicrobials for this application are formaldehyde and vaporized hydrogen peroxide (VHP). Both have been used for many years, with VHP becoming more popular due to the significant safety and efficacy concerns with the use of formaldehyde. The pharmaceutical industry has found that VHP is dry on contact, a rapid sporicidal agent, demonstrates broad spectrum material compatibility and is safe environmentally, as well as from a health risk exposure perspective. For example, hydrogen peroxide vapors have an 8-hour workday exposure of 80 ppm versus 0.1 ppm for ClO_2. Also note that in comparison to 500 ppm sporicidal concentration of ClO_2, VHP is sporicidal at 1–2 ppm at 25° C. It is important to note that this technology has been widely recognized and accepted. Major pharmaceutical companies such as Merck, Baxter, and Pfizer have used VHP technology as an effective room sterilizing agent for over 10 years. Further, the sterilant used in the VHP process has been registered with the EPA now for many years. Decontamination requires no hydration (in fact rooms are dehumidified to ~40 percent), is controlled, documented and can be validated parametrically, biologically and chemically, and is safe for use on a wide range of sensitive materials, including computers and electronics. For selected areas that have sensitive equipment or fine art, this technology may provide the only alternative with less damaging long-term effects. Additionally, if repeated treatment is required in highly contaminated areas to ensure eradication of spores without severe effects, vaporized hydrogen peroxide would be the fumigant of choice.

Currently, a single VHP delivery system can successfully decontaminate rooms up to ~7500 ft^3—and with multiple systems in tandem can decontaminate larger areas. A new, high capacity VHP delivery system has recently been designed and put on fast track development by STERIS to be available as soon as possible for larger scale room decontamination. A complete dossier of publications, successful application, case studies and user references are available on request.

We respectfully submit that reliance on a single technology may not be the most prudent course of action when attempting to significantly reduce the risk of infection due to this level and nature of contamination. We propose that independent verification of the discussions in this document be obtained from known thought leaders in sterilization and disinfection; for this we recommend Dr. Seymour Block, Department of Chemical Engineering at the University of Florida, a recognized expert in this area, as exemplified by being the editor of five editions of *Disinfection, Sterilization, and Preservation*.

APPENDIX A

TAB B

COMPARISON OF ANTIMICROBIAL FOGGING, CHLORINE DIOXIDE AND VHP FOR ROOM DECONTAMINATION

STERIS Corporation.—Comparison Tables

Fogging/foaming	Gaseous chlorine dioxide	VHP
Variable coverage and distribution	Depending on mode of delivery, more reliable distribution. Difficult to maintain in gaseous state; can condense.	Controlled delivery system for more reliable distribution. Kinetics of maintaining gaseous state is understood and important for process effectiveness.
Wet methods	Requires significant hydration for antimicrobial efficacy. Essentially wet process.	Dry sterilization method
High concentrations required for rapid sporicidal activity.	500ppm sporicidal over 8 hours	1–2ppm sporicidal at 25° C. 1 log reduction every 1–2 minutes.
Difficult to control and deliver over large surface areas and ensure residence time for horizontal surfaces.	Significant risks and difficultly providing to a large area. Overall better coverage than fogging or foaming.	Controlled delivery contacts all surfaces. New system available for large area fumigation
Difficult to validate	Validation possible. Can be biologically verified.	Validation and documentation routinely conducted. Can be parametrically, biologically and chemically verified.
Concerns over material compatibility; extent dependant on contact time and antimicrobial agent/formulation.	Significantly damaging to a variety of surfaces, even after single exposures. Concerns already noted in cleanroom applications.	Broad range material compatibility
Not safe on electrical equipment	Not safe on electrical equipment	Safe on electrical equipment
Efficacy inhibited by presence of absorbing materials.	Efficacy inhibited by presence of absorbing materials.	Efficacy inhibited by presence of absorbing materials
Occupational risks significant, dependant on antimicrobial used.	Occupational risks significant, but can be minimized.	Occupational risks minimal. Safest for environment and personnel health
Extended contact times and clean-up required.	Extended contact times and possible clean-up required. Chlorine residuals.	Most rapid method and room ready for use directly following cycle. No residuals.
Limited registration, depending on antimicrobial.	Unknown registration situation with process.	Sterilant used in the process registered with the EPA
Variable efficacy depending on the product.	Broad spectrum antimicrobial activity	Broad spectrum antimicrobial activity, including independant testing against and validation with B. anthracis.

For these reasons, VHP is highly recommended for the safe, efficacious decontamination of rooms as part of an overall strategy for building decontamination against B. anthracis spores and other potential bioterrorism microorganisms.

COMPARISON OF BLEACH AND REGISTERED SPORICIDAL PRODUCTS FOR LIQUID SURFACE DISINFECTION

STERIS Corporation.—Comparison Tables

Bleach/Sodium Hypochlorite	Registered Sporicidal Products
Slowly sporicidal, depending on active chlorine concentration. Data cited based on simple suspension study. In our hands, with purified spore suspension, we showed a 2–3 log reduction with B. subtilis spores within 15–20 minute contact.	Rapid and consistent sporicidal activity; for example, synergistic oxidizing agent formulations at low concentrations show a 6 log Bacillus spore reduction in <5 mins in similar suspension studies. Decreased time required at higher concentrations of the product. Enhanced efficacy due to synergy between actives.
Not registered sporicide or passed routine EPA requirements for antimicrobial efficacy.	Registered sporicide, virucide, bactericide and fungicide
Significant reduction in activity when spores present on a surface or in the presence of organic or inorganic soil (either on the contaminated surface or in the water used to dilute the product).	Demonstrated rapid activity on a surface and in the presence of soils.
Aggressive and damaging on surfaces, especially at sporicidal concentrations.	Broad range material compatibility, depending on product formulation
Not widely used for high risk surface decontamination	Widely used and validated in the pharmaceutical and research industries for antimicrobial, including sporicidal efficacy

For these reasons, registered sporicidal products would be a more desirable product for high risk contaminated areas.

COMPARISON OF ANTIMICROBIAL FOAM, LIQUID CHLORINE DIOXIDE AND REGISTERED SPORICIDAL PRODUCTS

Foam or nanoemulsions	Liquid Chlorine Dioxide	Registered Sporicidal Products
Very slow sporicidal activity, requires long exposure times.	Rapid sporicidal activity, depending on product claims.	Rapid sporicidal activity
Not registered, experimental technologies	Some registered	Registered, validated and widely used in regulated environments
Would require demonstrated broad spectrum efficacy demonstration by required standard methods for product claims.	Broad spectrum efficacy for those products with regulated claims.	Broad spectrum efficacy claims, including sporicidal, bactericidal and virucidal
Unknown material compatibility	Limited material compatibility	Broad range material compatibility under laboratory and in-use conditions/experience
Foam needs to 'release' active (broken) to be available for antimicrobial activity.	Liquid immediately available for instant antimicrobial activity.	Liquid immediately available for instant antimicrobial activity

APPENDIX A

TAB C

STERIS CORPORATION

FEATURES OF CURRENT PROPOSAL AND PROPOSED ALTERNATIVES

- B. anthracis—concentration of 10_{12} organism/gm
  - Infectious Dose 50 percent (ID_{50}) of 6,000–10,000 spores
  - Potential for cutaneous anthrax with as little as 5–50 spores in animal trials.
- Development of infection will also depend on the health and immunocompetence of an individual
- Building has contaminated and uncontaminated sections.
- No approach 100 percent effective for decontamination.
- No one method is capable of doing the entire decontamination process, but a logical series of steps can be taken to maximize the decontamination process
- Decontamination should be conducted with registered products/processes that are widely used for similar applications and will minimize damage to a wide variety of surfaces.

- **Building should be sealed and contamination mapped. High and low risk areas should be identified and interventions (either single or multiple) conducted to reduce infection risks associated with each area.**
- **Verification of process effectiveness by process monitoring and documentation**
- **Retesting for contamination following decontamination to confirm effectiveness.**

Draft Remediation under Consideration	Alternate Considerations
• Approach for disinfection HEPA vacuuming Bleach for hard surfaces Use of antimicrobial foam 1000 ppm of ClO_2 overnight or longer • Whole building, including non-porous materials, i.e. papers, drapes and rugs etc. to be treated by ClO_2 • ClO_2 believed to be compatible for short term exposure • Health considerations for ClO_2 deemed reasonable • Diluted bleach solution for hard surfaces has been deemed compatible	Drapes, papers and rugs will not be adequately disinfected by ClO_2 and may actually reduce the efficacy of room fumigation. Treatment of paper and drapes by ethylene oxide or irradiation is recommended, independent of whatever fumigation technique is adopted. Treatments should be considered and chosen to assure as low spore count as possible rapidly and safely on any surface. Registered products with proven efficacy for this purpose should be adopted. Diluted bleach will give variable results depending on the surfaces treated, will require long exposure times, will not be effective in certain conditions and will damage certain surfaces Treatment of carpet by strong sporicidal products directly can give higher assurances ClO_2 (liquid or gas) will damage surfaces such as metals and paints at the exposure time recommended Vapor H_2O_2 (VHP) applied in dry state will cause less damage and is widely accepted, safe and sterilant registered for this application.

APPENDIX B

COMPARISON OF ROOM DECONTAMINATION METHODS

Fogging/foaming	Formaldehyde	Gaseous chlorine dioxide	VHP
Variable coverage and distribution.	Variable coverage and distribution.	Depending on mode of delivery, more reliable distribution. Difficult to maintain in gaseous state; can condense.	Controlled delivery system for more reliable distribution. Kinetics of maintaining gaseous state is understood and important for process effectiveness.
Wet methods	Requires significant hydration for antimicrobial efficacy. Essentially wet process.	Requires significant hydration for antimicrobial efficacy. Essentially wet process.	Dry sterilization method
High concentrations required for rapid sporicidal activity.	High concentrations required for rapid sporicidal activity.	500ppm sporicidal over 8 hours (but needs to be kept in the dark).	1–2ppm sporicidal at 25° C. 1 log reduction every 1–2 minutes.
Difficult to control and deliver over large surface areas and ensure residence time for horizontal surfaces.	Significant risks and difficultly providing to a large area. Overall better coverage than fogging or foaming.	Significant risks and difficultly providing to a large area. Overall better coverage than fogging or foaming.	Controlled delivery contacts all surfaces. New system available for large area fumigation
Difficult to validate	Difficult to validate	Validation possible. Can be biologically verified.	Validation and documentation routinely conducted. Can be parametrically, biologically and chemically verified.
Concerns over material compatibility; extent dependant on contact time and antimicrobial agent/formulation.	Can be damaging to surfaces.	Significantly damaging to a variety of surfaces, even after single exposures. Concerns already noted in cleanroom applications.	Broad range material compatibility

Fogging/foaming	Formaldehyde	Gaseous chlorine dioxide	VHP
Not safe on electrical equipment.	Not safe on electrical equipment.	Not safe on electrical equipment.	Safe on electrical equipment
Efficacy inhibited by presence of absorbing materials.	Stable, difficult to remove ...	Efficacy inhibited by presence of absorbing materials.	Efficacy inhibited by presence of absorbing materials
Occupational risks significant, dependant on antimicrobial used.	Significant occupational and safety risks.	Occupational risks significant, but can be minimized.	Occupational risks minimal. Safest for environment and personnel health
Extended contact times and clean-up required.	Extended contact times and clean-up required.	Extended contact times and possible clean up required. Chlorine residuals.	Most rapid method and room ready for use directly following cycle. No residuals.
Limited registration, depending on antimicrobial..	Limited registration, traditional use.	Unknown registration situation with process.	Sterilant used in the process registered with the EPA
Variable efficacy depending on the product.	Variable efficacy	Broad spectrum antimicrobial activity.	Broad-spectrum antimicrobial activity, including independent testing against and validation with B. anthracis.

BATTELLE MEMORIAL INSTITUTE,
November 28, 2001.

MS. CAMERON TAYLOR,
Committee on Environment & Public Works,
U.S. Senate,
Washington, DC.

MS. TAYLOR: Per your request I have prepared brief summaries of some significant methods of decontamination. I have grouped these into Chemical Methods and Physical Methods. The Chemical Methods of decontamination are further divided into Alkylation and Oxidation methods, however, please keep in mind that in spite of decades of research, the exact mechanism of microbicidal action is often not well understood. Please recognize this is not an all-inclusive list of methods used for decontamination, disinfection, sanitization or sterilization, however, I believe I have covered most of the standards. Note certain of these methods, for example hypochlorites, actually represent a group of closely related chemicals. Note too that certain combinations of these fundamental methods provide synergy that is exploited for specific applications, for example hydrogen peroxide is used together with ultraviolet light, peracetic acid has been used in conjunction with plasma and so on. For confidentiality purposes I am not at liberty to include proprietary formulations, however, many, if not most of those, use permutations the basic methods listed here. For each method I have tried to address the fundamental considerations for a good decontaminant:

- Safety
- Efficacy
- Toxicity
- Economy
- Penetration capability
- Environmental impact
- Temperature of use
- Commercial availability

I may not have addressed capabilities of a specific methodology of interest to you, please let me know if there is more information I can provide. I look forward to hearing from you and hope you will allow me to respond to any questions you may have.

Respectfully,

C. DANIEL ROWE, PH.D., *Research Leader—Chemistry,*
Battelle Memorial Institute.

DECONTAMINATION METHODS—BACTERIAL SPORES

CHEMICAL METHODS—ALKYLATION

- Alkoxides—DS2
- Beta Propiolactone
- Ethylene oxide/Propylene oxide
- Formaldehyde /Formalin/Paraformaldehyde
- Glutaraldehyde
- Methyl Bromide
- Nanoparticles
- Phenolics
- Quaternary Ammonium Compounds

CHEMICAL METHODS—ALKYLATION

- Chlorine
- Chlorine dioxide
- ECASOL
- Fichlor
- Hydrogen peroxide

—Liquid
—Vapor

- Hypochlorite
- Iodine
- Peracetic acid
- Ozone

PHYSICAL METHODS

- Alcohols
- Plasma

Radiation

—Electron beam—Gamma—X-ray
—Ultraviolet

- Sorbents
- Steam

CHEMICAL METHODS—ALKYLATION

1. Alkoxides

Decontaminating Solution No. 2 (DS2).—Decontaminant Solution number 2 is the standard military decontaminant used on vehicles, equipment and building exteriors. It falls into the general class of decontaminants called "Alkoxides", where strongly alkaline materials (such as sodium hydroxide or potassium hydroxide) are dissolved into an organic solvent forming a very strongly basic solution. DS2 consists of 70 percent diethylenetriamine, 28 percent ethylene glycol monoethyl ether and 2 percent sodium hydroxide. DS2 requires 30 minute contact time with the surface being decontaminated and must be thoroughly rinsed off with water after decontamination. DS2 will neutralize many chemical and biological agents but does *NOT kill spores* (anthrax or otherwise), is corrosive to some metals, will soften leather and may soften remove or discolor paint. DS2 is toxic (protective masks, gloves and aprons must be worn during application). DS2 is flammable and will ignite of sprayed on surfaces over 168° F. DS2 is explosive when mixed with certain other decontaminants such as Super Tropical Bleach (cf. Hypochlorites). The military has over 2 million gallons of DS2 in inventory.

2. Beta Propiolactone

Beta propiolactone is an organic liquid that is used as an intermediate in chemical synthesis. It is also used to sterilize vaccines, grafts and plasma. It has been proposed as a substitute for formaldehyde in decontamination of rooms and buildings. Research has shown beta-propiolactone to be 1000 times more active than ethylene oxide and 25 times more effective than formaldehyde, however it is not recommended as a substitute for ethylene oxide because it does not have the penetrating power of ethylene oxide. It has been shown to cause cancer in mice.

3. Ethylene Oxide / Propylene Oxide

Ethylene oxide (EtO) is a colorless, flammable, toxic gas at room temperature (boiling point 51 degrees Fahrenheit). EtO has been used for the past 50 years as the principal method for sterilizing heat sensitive equipment (e.g. surgical equipment) and as an agricultural fungicide and fumigant for foodstuffs and textiles. To reduce the explosion hazard EtO is often mixed with a nonflammable, nontoxic gas such as carbon dioxide or CFC's (fluorocarbons). EtO causes cancer and mutations, is highly irritating to the eyes and can cause pulmonary edema. It is normally employed inside a closed pressure vessel specifically designed for EtO sterilization. Because of its low boiling point, EtO is used at high concentrations (300 to 1,200 mg/L) and thereby has excellent penetrating capabilities. EtO is noncorrosive and highly sporicidal. A major disadvantage of EtO is that it dissolves into materials such as plastics and hence requires long aeration periods (e.g. 12 hours) to eliminate the residual gas. Also EtO is only effective above specific humidity levels.

Propylene oxide is similar to EtO but is a liquid at room temperature (boiling point 95° F is higher than EtO) and is used more often on foodstuffs in the liquid form.

4. Formaldehyde / Formalin / Paraformaldehyde

Formaldehyde is a colorless, flammable gas (boiling point minus 3 degrees F) at room temperature. Formaldehyde is used as a germicide in either the gaseous or liquid state and is a potent respiratory irritant. Formaldehyde is toxic and has a pungent suffocating odor that is intensely irritating to mucous membranes and eyes and causes contact dermatitis, violent coughing, and death if ingested. Formaldehyde is known to cause cancer. OSHA has established employee exposure limits of 0.75 ppm for 8 hour time-weighted average exposures and 2 ppm exposure for 15 minutes (short-term exposure limit). Formaldehyde readily polymerizes with itself thus leaving a sticky residue. It also readily condenses on cold surfaces such as doors making it difficult to control the exact desired concentration. Formaldehyde does not penetrate into porous surfaces, fissures or device lumens. It is noncorrosive, however, it does require >70 percent humidity to be an effective biocide. The preferred contact time is 18 to 24 hours. Explosions occur when formaldehyde concentrations exceed 7 percent.

Formaldehyde gas is employed in specially designed pressure vessels as a medical device sterilant (Europe). Formaldehyde gas is also produced by heating formalin (aqueous solution of formaldehyde gas) or paraformaldehyde (a white powder that is a polymerized form of formaldehyde). A quantity of the powder is placed in a frying pan and heated. It is used to disinfect sickrooms, clothing, linen and sickroom utensils. Paraformaldehyde (formaldehyde) is sporicidal and is routinely used in Fort Detrick, MD to decontaminate buildings in which anthrax has been used. Sensitive equipment and books etc. are required to be removed from the room or building prior to decontamination. After decontamination using formaldehyde several days are required for aeration and washing (walls etc.) using copious amounts of water to eliminate odor and toxic residue. Residual formaldehyde levels can be very high in polyester, rubber and cellulosics (paper, cotton, wood etc.). Disposal of formaldehyde must be completed in accordance with regulations for toxic waste.

5. Glutaraldehyde

Glutaraldehyde is "double aldehyde" and a chemical relative of formaldehyde except it is used in an aqueous solution. Like formaldehyde, glutaraldehyde has irritating odors and toxic fumes. Glutaraldehyde is sold widely in 2 percent aqueous solutions for use in hospital instrument disinfection and surface decontamination for temperature sensitive instruments that cannot be steam sterilized. It is known to cause skin and eye irritation and causes some people to be extremely sensitive to its vapors as it builds in their system (it is used as an embalming fluid at 25 percent). It is relatively noncorrosive, however it's effectiveness is strongly influenced by pH, temperature and concentration. For example, at high pH and high temperature (120° F) glutaraldehyde's efficacy is poor, likewise at low pH and low temperature (65° F) it also has poor efficacy. Glutaraldehyde (2 percent) rapidly kills some organisms, however, to be sporicidal it requires 10 to 14 hours contact time. Rinsing is required after glutaraldehyde decontamination. Glutaraldehyde penetrates device crevices and can be difficult to remove because it produces an intractable polymeric residue.

6. Methyl Bromide

Methyl bromide is a colorless, odorless, extremely toxic gas at room temperature. It has been used to fumigate poultry houses, warehouses, vaults, mills and freight cars. It has also been used to extract oils from nuts, seeds and flowers, however residues have raised questions about safety aspects for consumers. Methyl bromide has a greater penetrating power than formaldehyde and is more easily diffused from buildings than formaldehyde. Although Methyl bromide is effective against a wide range of bacteria and viruses, I do not know the efficacy against spores such as anthrax. High concentrations are fatal to humans; chronic exposure can cause depression of the central nervous system and kidney failure. Due to its extreme toxicity, there has been great concern over distributing this gas in small fire extinguishers.

7. Nanoparticles

Nanoparticles refer to compounds that have been made in such a way that they have unusually small particle sizes. Even though they are solid materials the sizes of the solid particles are in the molecular range (0.000,000,15 inch). This has the effect of enormously increasing the surface area of the bulk material and changing the chemical and physical properties. (Seven grams of nanoparticles have the surface area equivalent to a football field). Although there appears to be a great deal of research underway on nanoparticles two approaches (that I know of) have been proposed for decontamination.

• Metal oxides: By manufacturing common metal oxides (for example magnesium oxide) using an aerogel process, it exhibits properties it would not normally exhibit. These ultra small particle sized metal oxide particles have been doped with chlorine or bromine and demonstrated to kill some spores under certain conditions within 1 hour. A disadvantage of this method is that the material is a finely divided powder and as such is extremely difficult to remove from some sensitive equipment, fabric, machinery, joints and crevices.

• Nanoemulsions: A second approach to nanoparticles is called "Nanoemulsions". This technology employs a fat/oil phase (e.g. soybean oil) with a water phase containing detergent and other additives blended using a great deal of force (shear forces). The result is an extremely small oil droplet that is able to react with and disrupt the membrane of bacteria and hence exhibit antimicrobial activity.

These emulsions require up to 4 hours to achieve complete spore kill. Nanoemulsions would be difficult to use for buildings due to the residue remaining.

8. Phenolics

The organic liquid, phenol or carbolic acid is one of the oldest germicidal agents used in the hospital environment. The parent chemical has been replaced by hundreds of derivative compounds referred to as phenol derivatives or phenolics (an example of this is resorcinol). Phenol no longer plays a significant role as an antibacterial agent. These are considered to be moderate to low level disinfectants used for environmental surfaces and non-critical devices in institutional and commercial environments. Phenol derivatives are also used as preservatives and antibacterial agents in germicidal soaps and lotions. Since small amounts of phenolics (1 percent to 2 percent) remain active when in contact with gross amounts of organic soil, phenolics are often the disinfectant of choice for general housekeeping and laboratory disinfection. Phenolics are difficult to rinse from most materials and residues may irritate skin, cause depigmentation and cause injury to mucous membranes. Phenolics are not sporicidal.

9. Quaternary ammonium compounds

A wide variety of quaternary ammonium compounds (e.g. benzalkonium chloride) have been used since their introduction as liquid chemical germicides in 1935. A great deal of study has gone into examining the relationship between chemical structure and biocidal properties. These compounds vary greatly in their structure and likewise vary in their water/oil solubility and hence their microbicidal properties. The biocidal activity is derived, in part, from the surfaceactive (surfactant) nature of these compounds. These compounds find extensive application in consumer products but in most cases do not appear to exhibit pronounced sporicidal activity.

CHEMICAL METHODS—OXIDATION

1. Chlorine

Chlorine is a reactive, greenish-yellow gas with a suffocating odor that is dangerous to inhale. Chlorine readily reacts with almost all chemical elements (except noble gases, carbon and nitrogen). It is used for bleaching, purifying and disinfecting water and is extremely valuable as a chemical intermediate for hundreds of materials (plastics, rubber, chlorinated organic chemicals). Gaseous chlorine was used as a chemical weapon in World War I. Chlorine gas exhibits rapid biocidal activity. Chlorine, in aqueous solution, even in minute amounts, exhibits fast bactericidal action. Early research showed that aqueous solutions of chlorine get their biocidal activity by the formation of hypochiorous acid (cf. Fichlor and ECASOL).

2. Chlorine Dioxide

Chlorine dioxide is a gaseous oxidizing agent that has been used for many years as a drinking water disinfectant and as a bleaching agent for paperpulp, flour, leather, fats, oils, textiles and beeswax. It is unstable at higher concentrations so, like ozone, it must be generated onsite. Its solutions can emit toxic and corrosive fumes that require its use to be restricted to locations having proper engineering controls. It is reacts violently with organic chemicals and is unstable in sunlight, 10 percent concentrations can easily be detonated by sunlight, electrical discharge or decompose explosively. Chlorine dioxide is a more severe respiratory irritant than chlorine and a 19 ppm concentration has reportedly caused occupational fatality. Inhalation exposure symptoms include coughing, wheezing, runny nose, eye and throat irritation, headache, vomiting, bronchitis and pulmonary edema (a life-threatening accumulation of fluid in the lungs). Pulmonary edema may be delayed for several hours after exposure. Chlorine dioxide also may ignite combustible materials for example mixtures with fuels may explode and chlorine dioxide containers may explode when exposed to heat, fire, friction or contamination.

Sewer run-off may also create a fire or explosion hazard. Bacterial spores have not been the primary test organisms since spores are not generally regarded as important waterborne pathogens consequently the efficacy of chlorine dioxide against anthrax spores may need to be studied.

3. ECASOL

Electrochemically activated solution (ECASOL) is produced by applying a short-term (0.3–0.6 sec) mild electrical charge (5–9 amps at 8–12 volts) to a dilute brine solution (less than 1 percent salt) as it passes through a flow-through electrolytic module (FEM). Electrolysis occurs (i.e. separation of the ions using an electric current) thus producing a highly oxidative solution. The primary active ingredient of the solution is hypochlorous acid. Because hypochlorous acid at the applied concentration is a weak acid (neutral pH) it is non-corrosive and safe for skin, eyes and wounds. In addition to being a weak acid and skin-safe, hypochlorous acid is an oxidant so powerful it is an effective biocide even at low concentrations. The biocidal efficacy of hypochlorous acid (pH 7) is many times greater than the biocidal efficacy of an equivalent amount of its chemical relative, hypochlorite (pH 10). ECASOL at 0.035 percent (350ppm active chlorine) kills 100 million anthrax spores on a test coupon instantaneously. The active ingredients in ECASOL are relatively unstable and decompose readily in the environment reverting back to the starting materials of salt and water.

4. Fichlor

Fichlor is a commercially available oxidant (sodium d ichloroisocyanu rate). It is used as a sanitizing and disinfecting agent in commercial bakeries and as a bleaching agent, swimming pool disinfectant, dishwashing detergent and in cleaner formulations. It is believed that, in aqueous solutions, the reactive species is hypochlorous acid. Fichlor dissolves in water at the same rate as calcium hypochlorite and is used similarly to calcium hypochlorite.

5. Hydrogen Peroxide

Liquid Hydrogen Peroxide.—Hydrogen peroxide is a noncarcinogenic, nonmutagenic nontoxic liquid oxidant that decomposes to environmentally safe, residue-free by-products (water and oxygen). It is widely used for bleaching, aseptic food packaging, dairy processing, medical device sterilization (e.g. hemodializers, pharmaceutical preparation areas) and bacterial reduction and odor control for sewage effluent. Hydrogen peroxide solutions have been used as surface disinfectants for 150 years. Aqueous hydrogen peroxide solutions in relatively high concentrations (10–30 percent) are observed to have sporicidal capability (note >50 percent peroxide solutions are extremely dangerous and can be unstable). Lower concentrations (e.g. 6 percent) require lengthy contact times (30 minutes or greater) in order to be sporicidal. Hydrogen peroxide has some material compatibility problems at high concentrations (nylon embrittlement, oxidizes aluminum and discolors anodized aluminum etc.). Hydrogen peroxide requires long contact times in order to kill resistant bacterial spores. Hydrogen peroxide is the primary active ingredient of the Sandia Foam decontaminant.

Vaporized Hydrogen Peroxide.—Cold gaseous hydrogen peroxide is known to be a sporicide at low concentrations. Hydrogen peroxide is a far more effective sporicide in the vaporized form than it is in liquid solutions (note this is not a mist or a fog but a dry gaseous vapor). It exhibits a broad range of microbicidal activity over a broader temperature range than ethylene oxide having been shown to be an effective sporicide at temperatures as low as 4 degrees C and as high as 80 degrees C.

The sporicidal efficacy of gaseous hydrogen peroxide improves with increasing concentration, increasing temperature and increased exposure time. Greatest lethality is achieved at near saturation levels. It is effective on exposed, clean, dry surfaces when there are no contaminants such as liquids, oils, salts or organic residues to impede penetration. In other words it is an effective surface decontaminant and does not have the penetration capability even close that of ethylene oxide. Note cold surfaces readily condense vaporous hydrogen peroxide and hence greatly reduce the peroxide available for decontamination. Vaporized hydrogen peroxide is rapidly, efficiently and cleanly decomposed by metal catalysts (e.g. platinum) to water and oxygen. Decontamination can be achieved in relatively short contact times; thus, minimizing concern over material effects on items being processed. Because of the sporicidal nature of vaporized peroxide at low concentrations (<10mg/L), at room temperature (<80 degrees F) and because of the inherently low toxicity of hydrogen peroxide and its by-products it is employed as a non-toxic cold gas sterilant for medical devices.

6. Hypochlorite compounds

Calcium hypochlorite is a powerful oxidizing agent and bleaching agent. Aqueous hypochlorite solutions have high pH (alkaline) and hence have hydroxide ion present as well and are thus caustic and corrosive. This limits its biocidal effectiveness and adversely effects material compatibility. Hypochlorite solutions are made by mixing dry calcium hypochlorite and water. The military employs 0.5 percent bleach solutions as personnel decontaminant in spite of the fact that it is highly irritating to the skin and cannot be used in eyes or wounds. Note sodium (as opposed to calcium) hypochlorite is sold commercially as a 5 percent solution (Clorox®) but sodium hypochlorite is not stable as a dry solid. The military employs HTH (high test hypochlorite) and STB (super tropical bleach) Hypochlorite solutions can ignite spontaneously on contact with DS2. Solutions of 0.5 percent calcium hypochlorite (2800 ppm active chlorine) kill one hundred million anthrax spores on a test coupon in 30 minutes. Hypochlorite solutions in excess of 1 percent create toxic chlorine gas vapors.

7. Iodine

Iodine is a grayish-black solid that looks like metallic scales. It is very slightly soluble in water and forms a brown solution. Compared to chlorine the chemistry in water and activity is much more complex. Iodine reacts not only with living organisms but dead ones and dissolved proteins. Iodine is able to penetrate the cell wall of a microorganism rapidly. Many disinfectant formulations employ iodine or iodophors. lodophors are a combination of elemental iodine or triiodide with a carrier. Here the iodine is stabilized by an appropriate surfactant (surface active agent). Iodine itself is bactericidal, fungicidal and *sporicidal*. It is most active as acid pH (i.e. low pH). lodophors (iodine carriers) retain the sporicidal strength but not the undesirable properties of iodine and are active over a wide temperature range.

8. Peracetic acid

Peracetic acid is essentially vinegar (acetic acid) with an extra oxygen atom. Peracetic acid has been known as a germicide for almost 100 years. It is a powerful oxidant, has an acrid odor, is corrosive to the skin and explodes violently when heated to high temperatures. Dilute aqueous peracetic acid solutions are used in the food industry as a disinfectant spray (dairy industry, fruits, vegetables for mold growth etc.). Prolonged exposure with peracetic acid will damage most materials and it will cause burns and blisters on the skin. It is a strong acid and as such is very corrosive to metals; however, buffered solutions containing anticorrosion additives are widely used (0.2 percent concentration) as a sterilant for surgical devices. *Liquid* peracetic acid has been used for years to reprocess kidney dialyzers and to sterilize immersible surgical instruments whereas *vaporized* peracetic acid has been used in combination with plasma for medical device sterilization and is used by the CDC as a high level disinfectant.

9. Ozone

Ozone is triatomic oxygen and is a pale blue gas (boiling point minus 170 degrees Fahrenheit) that is relatively unstable. The half-life of gaseous ozone in ambient atmosphere is 12 hours and the half-life of aqueous ozone is 30 minutes. It is the most reactive form of oxygen and has a very high oxidation potential second only to fluorine therefore it is one of the strongest oxidants known. Ozone must be generated onsite and is used to disinfect drinking water, reduce odors in the paint industry, as a bleaching agent and many other applications. High concentrations are injurious and cause severe irritation of the respiratory tract. Generally, ozone is known to be a more effective bactericide and virucide than chlorine and chlorine dioxide. Because of the highly oxidative properties of ozone, it appears to be best used for decontaminating items composed of silicone rubber, ceramics, polyvinyl chloride, polyurethane or metals such as titanium, stainless steel, platinum and other metals inert to reaction with oxygen.

PHYSICAL METHODS

1. Alcohols

The alcohols have been appreciated for centuries for their antiseptic qualities. As a chemical group, the alcohols possess many features desirable for a decontaminant. They are bactericidal against non-spore forms, are relatively inexpensive, easily obtained, nontoxic, have a cleansing action and evaporate readily. They are much less powerful against spores than they are against vegetative organisms. Alcohols are believed to get their biocidal activity by denaturing proteins. This is shown to be far more effective in the presence of water, hence water and alcohol mixtures have much greater biocidal activity than absolute alcohol. Methyl alcohol (wood alcohol) is the weakest of bactericidal action of the alcohols and is seldom considered for use as an antibacterial agent. On the contrary ethanol (ethyl alcohol) when water is present is a very effective bactericide against vegetative organisms; however it has little effect against bacterial spores. The bactericidal action of isopropyl alcohol is slightly greater than that of ethanol. The inability of alcohols to destroy bacterial spores makes their use as a decontaminant inadvisable.

2. Plasma

Plasma is sometimes referred to as a "fourth" state of matter (as compared to solid, liquid and gas). A majority of all matter in the universe exists as plasma. Plasma is best thought of as a high energy "soup" consisting of a mixture of atomic fragments including positive ions, negative ions, free electrons and free radicals (no charge). Plasma is a good conductor of electricity and is affected by magnetic fields. The light or "glow" from plasma occurs as electrons reassume their positions (i.e. decay) in the atomic orbitals from whence they came and in so doing, releasing a photon of light. Examples of plasma include neon lights, the northern lights, fluorescent lights, lightning and many stars. Industrial plasmas are used for their destructive properties on materials, for example, the auto industry uses plasma to strip polypropylene surfaces and thereby improve paint adhesion to the surface. The semiconductor industry uses plasma for etching surfaces. Although low temperature plasmas have been claimed to serve as stand-alone method of decontamination, in no case has plasma alone been used without having an additional (chemical) decontaminant present. For example, one medical device sterilizer uses plasma along with vaporized hydrogen peroxide as the sterilant, another uses vaporized peracetic acid as the sterilant. As a result there is not a strong argument to use plasma as a stand alone decontaminant. Plasma suffers from the disadvantage of not having great penetrating power, will react with paper and other materials, is inactivated by highly absorptive materials, is unable to enter small deep lumens (i.e. passages) and is significantly reduced in efficiency by the presence of blood or salt crystals.

3. Radiation

Radiation can be classified into two main groups: electromagnetic (e.g. ultraviolet, gamma, x-rays and microwave) and particle radiation (alpha, beta, neutrons etc.). Radiation produces bactericidal effects by transferring the beam energy into ionization of the biologic target (except in the case of ultraviolet). This effect is usually produced with no perceptible rise in temperature and is therefore referred to as ionizing radiation for "cold sterilization". Greater than 50 percent of medical devices sold sterile are irradiated (especially high volume disposables); radiation is also used for food sterilization.

Electron beam (radiation).—Electron beam radiation or "E-beam" radiation is a stream of high-energy electrons (i.e. beta particles) accelerated to a high energy in a radio frequency linear accelerator. E-beam radiation is an alternative to ethylene oxide and gamma sterilization. E-beam sterilization is accomplished by exposing a product for a predetermined time to the high-energy electron beam. Exposure to E-beam disrupts the bonds of vital metabolic components bonds in a microorganism, rendering the exposed product sterile. In order to achieve adequate penetration the electrons must be accelerated to high energies, consequently E-beams are usually large expensive machines and generally require massive shielding capabilities to protect personnel. To provide a shield from the energy produced, products are often processed in a concrete bunker. More recent models have been developed that are smaller and have more compact shielding. E-beam is the method currently being used to decontaminate mail from anthrax. The use of E-beam is limited due to limited ability of the E-beam to penetrate various materials and related dosimetry problems. E-beam is usually most effective when the load put through the beam is very uniform and hence the dose is also uniform. Throughput rates (dose) are determined by the beam energy and conveyor speed hence it has the advantage of being able to be turned on and off (unlike radioactive sources of gamma radiation). E-beam will severely embrittle and hence crack some plastics such as polypropylene; it will harden polyethylene and can discolor some materials such as cellulosics (wood, paper, cotton).

Gamma (radiation).—Gamma radiation is electromagnetic radiation generated from a radioactive material. Gamma radiation is a method of commercial sterilization accomplished by exposing products to cobalt-60, an isotope that emits gamma radiation. Gamma-sterilization is a one-step process that does not require any preconditioning or post-processing treatment of the product. (During the radiation process the packaged products are loaded onto conveyor system that transport products through an irradiation chamber or cell.) While cobalt-60 emits radioactive energy, this energy does not cause exposed substances to become radioactive, as a result, sterilized products have no residual radioactivity and can be shipped safely to customers immediately after processing. The dose applied to the product is determined by the amount of cobalt-60, the distance from the radiation source and the duration of the exposure to the radiation source. Products are typically processed using gamma radiation in 8 hours or less. The use of gamma radiation as a sterilization method is limited by the radiation compatibility of certain plastics and other materials that may discolor, deform or become brittle when exposed to gamma radiation. The use of gamma sterilization has increased in the past 10–15 years as medical products manufacturers have converted sterilization from ethylene oxide to gamma sterilization and have increasingly used radiation compatible materials in the new products. Gamma radiation has low environmental impact under normal operating conditions since there are no chemical residuals on the sterilized products or emission released by either the process or the sterilized product. Gamma rays have an advantage over E-beam in that gamma has generally better penetration power, however it suffers from the major disadvantage that radioactive sources cannot be turned on and off and require periodic replenishment due to half-life decay (cobalt-60 decays at 12 percent per year hence has a half life of 5.6 years). Gamma radiation also requires costly disposal of the spent cobalt-60. Potential liabilities associated with gamma radiation include worker exposure and radioactive contamination resulting from the use of cobalt-60.

X-ray (radiation).—X-rays are like gamma rays in that they are both electromagnetic waves. X-rays differ from gamma rays in that they are different wavelengths (hence different energy). The mechanism for decontamination with x-rays and gamma rays is the same for both. The difference is that different energy levels penetrate materials at different levels.

Ultraviolet (radiation).—Ultraviolet radiation is called "nonionizing" radiation because it does not cause ionization (i.e. it does not cause the electrons to be excited out of their orbital shells). Ultraviolet radiation has been proven to be a very effective biocide and is widely used for purification of air and water. The most practical method for generating UV radiation is by passing electrical discharge through mercury vapor in special glass tubes. Special glass tubes are required because ultraviolet light will not pass through most glass. The fluorescent lamp operates on the same basic principle producing ultraviolet energy. The difference between the two is that the fluorescent lamp is coated with a phosphor that converts the UV energy into visible light (glass used in most fluorescent lamps filters out all germicidal ultraviolet). The germicidal ultraviolet lamp is not coated with the phosphor and uses a glass that permits UV to pass. Since living organisms have the ability to repair some damage due to ultraviolet radiation, germicidal lamps must provide enough energy to exceed the tolerance limits to create lethal effects. Experiments have been done demonstrating whole room decontamination using several germicidal lamps employed simultaneously.

4. Sorbents

Sorbents are usually finely divided (small particle) powders that consist of essentially inert materials such as carbon, silicon dioxide, aluminum oxide, diatomaceous earth, kaolin, soil that adsorb chemicals on the sorbent surface. Sorbents, in general, do not react chemically to neutralize a contaminant but adsorb it on the surface so it can be physically removed. Sorbents can be made that will physically absorb many times their weight in specific chemicals. Sometimes the sorption process can be reversed by heating the sorbent in order to release the adsorbed chemical; on the other hand sometimes adsorption is irreversible. Sorbents can be chemically modified (e.g. polymeric sorbents) to make them chemically reactive toward specific chemicals (i.e. a *chemical* reaction as opposed to the normal sorption process which is *physical*). Sorbents are more often used for removal of chemical agents rather than for biological organisms.

5. Steam

Steam sterilizers or autoclaves have been made in America for over a hundred years. Steam sterilization is perhaps the most widely accepted method for decontamination and has been extensively studied; the mechanism of microbial inactivation using heat is fairly well understood. Since World War II increasing use of plastics and heat sensitive materials has spurred development of numerous low temperature decontamination methods especially in the medical device arena. Nevertheless, steam sterilization continues to be a workhorse for hospital sterilization of non-heat sensitive instruments and has found limited application in decontamination of infectious waste.

○